T0304101

Routledge Revivals

States' Finances in India

First published in 1968, *States' Finances in India* not only provides the first descriptive account of the finances of the states since Indian Independence, but also offers an analysis which successfully puts into perspective national, state and local finance on the one hand and problems of plan financing, taxation, borrowing and public expenditure on the other. The author supports his argument by a series of statistical tables on which he comments concisely, bringing out clearly their main features and implications. His analysis benefits from his practical experience of politics and administration.

He deals with union-state relations, state-local relations and problems of plan financing and implementation. Most of the statistical tables are aggregated from successive annual reports and will make this a valuable work of reference for economists, administrators, and politicians.

States' Finances in India

A Perspective Study for the Plan Periods

K. Venkataraman

First published in 1968
by George Allen & Unwin Ltd.

This edition first published in 2024 by Routledge
4 Park Square, Milton Park, Abingdon, Oxon, OX14 4RN

and by Routledge
605 Third Avenue, New York, NY 10017

Routledge is an imprint of the Taylor & Francis Group, an informa business

Publisher's Note
The publisher has gone to great lengths to ensure the quality of this reprint but points out that some imperfections in the original copies may be apparent.

Disclaimer
The publisher has made every effort to trace copyright holders and welcomes correspondence from those they have been unable to contact.

A Library of Congress record exists under LCCN: 68091605

ISBN: 978-1-032-84197-7 (hbk)
ISBN: 978-1-003-51164-9 (ebk)
ISBN: 978-1-032-84198-4 (pbk)

Book DOI 10.4324/9781003511649

STATES' FINANCES IN INDIA

A PERSPECTIVE STUDY FOR THE PLAN PERIODS

by

K. VENKATARAMAN

London

GEORGE ALLEN AND UNWIN LTD

RUSKIN HOUSE MUSEUM STREET

FIRST PUBLISHED IN 1968

PRINTED IN GREAT BRITAIN
in 10 on 11 point Times Roman type
BY WILLMER BROTHERS LIMITED
BIRKENHEAD

For my parents

PREFACE

There are today many fields of social and governmental activity where continually changing situations demand restatement of problems and redefinition of perspectives. No neat solutions await such problems and one can only hope that perspective analyses will lead to better functional understanding and 'working' solutions. Many major governmental problems today have political, administrative and academic implications and these have to be superimposed on each other, as in colour printing, for proper understanding.

It was the realization of the need for perspective analyses that impelled me to write *Local Finance in Perspective* three years ago and to attempt a work on *States' Finances in India* now. What started primarily as a perspective study had to play the part of a descriptive work as well, since there are no books on the finances of the State Governments in India after Independence. The result, I hope, is not unjust to either. But readers will notice that I have had to repeat the central themes and trends in some places for the sake of perspective; present considerable statistical material but at the same time eschew its detailed analysis for the same reason; and avoid tempting excursions into theory. The non-specialist reader may get bored with the many tables but I have considered them necessary since most of the material is not available elsewhere in a 'ready made' form.

The major portion of the work was written between two spells of arduous official work. The compilation of the statistical material also involved a great deal of labour. The credit for seeing me through the work should go to Padma, my wife, and my mother, Mrs Chellammal and above all to my father, Mr M. K. Krishnaswamy Iyer, who is a great beneficient influence in my life.

To the Madras Government, to whose cadre of the Indian Administrative Service I belong, I owe my gratitude for according me permission to publish the work. Needless to say, the views and the mistakes are entirely mine.

Madras
May, 1967

K. VENKATARAMAN

CONTENTS

Note: **1 crore = 10 million = 100 lakhs.**

LIST OF TABLES

LIST OF TABLES—*Continued*

CHAPTER I

THE SUBJECT

What makes the study of the finances of the state governments of India worthwhile is not only that the state governments play a pivotal role in the economic development of India but that there are various aspects of their finances, which, though doubtless requiring study, have not been so studied. In a sense, a correct perspective of state finances is not easily available except to those who are intimately connected with it. Probably the one major aspect which has received frequent and competent treatment has been the federal financial aspect. The principles and manner of distribution of resources between the Union and the States and the financial aspects of Union-State relations in general have been keenly watched and commented upon. The deliberations of the Finance Commission once in five years have provided a convenient opportunity for a study of state finances in general but here too the federal financial aspects have stolen the limelight. The formulation of each Five Year Plan has provided yet another occasion when the role of the states in the implementation of the Five Year Plans in the financial as well as administrative sense could be assessed. But plan documents cover a very wide field and cannot be expected to provide a delineation of state finances in any great detail. At the moment of writing the Administrative Reforms Commission is also addressing itself to an assessment of Union State relations in various spheres. Again, the emergence of non-Congress Governments in a number of states as a result of the Fourth General Elections has caused a lively debate on the future of Union-State relations.

While the federal financial aspects of state finances are no doubt very important, a discussion which centres on these aspects alone is apt to neglect the intrinsic financial structure of the states and also their position and performance inter se. As it is, even a 'straightforward' factual account of state finances since Independence is lacking. There are accounts of the finances of a few individual state governments but there is neither a comprehensive nor a comparative study of this important aspect of Public Finance.[1] Since 1950, the Reserve Bank of India Bulletin has been publishing annually a summary article on the finances of state governments. But these articles are confined to a description of the position of the state governments as revealed in their annual budgets. Probably because of the difficulties in comparison between different periods on account of problems of classification and presentation as well as the reorganization of States, the Bulletin has not so far seriously attempted a comparative study of state finances.[2]

There are thus considerable gaps in our knowledge, let alone perspective, of state finances.[3] There is a wide range and variety of questions which a student of public finance would find it worthwhile to ask. What part have state finances played in the implementation of plans and what effect have the plans had in turn on the finances of states? What is the nature and direction of federal assistance? What is the constitutional position of the states in the federal set up and what is the actual position and how far have financial trends acted upon constitutional provisions? What are the political and economic implications of the changing constitutional position? What part have state governments played in mobilizing resources for development? What are the trends and directions of revenue and expenditure? How far have the state governments discharged the functions cast on them by the Constitution? How far have regional disparities been reduced?

These are far too many questions to answer and surely they cannot be treated adequately in a work of this size. However, it is worthwhile to pose some of the questions in our attempt to provide a brief but analytical account of the states' finances since Independence and particularly since the First Five Year Plan. It is unnecessary in this connection to emphasize the important role played by state governments in the economic development of the country. Table 1.1 shows the size of the State Plans in relation to the total Plans.

[1]e.g. Dibakar Jha, Bihar Finances, 1912–13—1960–61. (Granthmala Karyalaya, Patna), B. V. Narayanaswami Naidu's account of Madras finances is before the era of planning. Goswami's Economic Development of Assam (Asia Publishing House) has only an appendix on the finances of Assam and V. V. Ramanadham's Economic Development of Andhra Pradesh (Asia Publishing House) has no discussion on finances.

[2]One notes with pleasure that the Reserve Bank of India Bulletin has attempted a study of 'State Governments' Expenditure 1951–52—1965–66' in its issue of June, 1966.

[3]We shall not be dealing with Nagaland and Union territories. We shall also ignore the bifurcation of Punjab into Punjab and Haryana States in October, 1966.

TABLE 1.1

Share of State Plans in Five Year Plans

(Rs. Crores)

Plan	Public Sector Outlay	States' Plan Outlay	Percentage of (3) to (2)
(1)	(2)	(3)	(4)
First Plan (1951-52 to 1955-56)	1,960	1,427	73%
Second Plan (1956-57 to 1960-61)	4,600	1,981	43%
Third Plan (1961–62 to 1965–66) (estimated)	8,631	4,155	48%
Fourth Plan (1966–67 to 1970–71) (estimated)	16,000	7,073	44%

[Source: 1. Third Five Year Plan, Pp. 88–89.
2. Fourth Five Year Plan, a Draft Outline, 1966, p. 74.]

If one reckons central expenditure on essentially state subjects (like education) the percentage will be even more.

But the states' role is vital not only in quantitative terms but in qualitative terms. It is the state governments that are nearer to the common man. Delhi is distant physically and figuratively. In a country of this size administration cannot function and planning cannot reach the people without the state governments. The climate for economic development is actually provided by them. The important responsibilities of the states in the maintenance of Law and Order and in Education and Public Health have necessarily to be well discharged if development is to be possible. The provision of the social overheads of development is their essential responsibility. Nor can challenging tasks of agriculture be faced without the agency of the states which are nearer to the land and the people. Irrigation and Power so essential to the common man as well as the country at large are again state responsibilities. Even for large public sector undertakings which the central government establishes, it is the state government that has to supply the land, water and electricity necessary.

To put it again in different words, 'the Centre builds up and maintains the overall instrumentalities of national economic life such as the credit and the monetary system, the railways and posts. It also acts in relation to the basic requirements of the long-term plan of industrialization with emphasis on large industry and exploitation of mineral resources. The states are concerned on the other hand with acting on the total life of all the people in their charge and on all the diffused, dispersed small scale units and activities. The Centre is concerned with the most generalized features of the national frame and with highly concentrated action at a few strategic points; the states must affect all areas and localities, all the relevant fields and all units. The Centre is concerned with the strategy of the long-term plan and with initiating the crucial movements, the states have to engage themselves in transmitting the forces impelling economic development to all areas and units and with concretizing for the individual units the fruits of development. The generalized objectives of a state plan are therefore making possible, initiating and encouraging economic development in all activities and sectors and areas and localities and protecting the standard of living and improving and ameliorating the situation, social and economic of all individuals within their territories. The locality and the individual are placed at the centre of the activities of states and providing for the universal impact of the developmental process and for a diffusion of its efforts becomes their primary aim'.[1]

As a matter of fact, the states themselves are in some respects relatively unwieldy units for the effective implementation of the Plans. The states in India vary in size and population and it would be an interesting study to examine the consequences of the size of the states, that is to say, how far particular states have been helped or handicapped by their size. More importantly state governments have themselves found that in essential tasks of development in the agricultural and rural fields and in reaching the man in the village, it has been necessary to decentralize the responsibilities and depend on Panchayati Raj institutions. In a country like India, local bodies have an essential role to play in economic development and to this extent state governments have, to varying degrees, delegated func-

[1]D. R. Gadgil, *Planning and Economic Policy in India* (Asia Publishing House, 1965), pp. 237–8.

tions and finances to local bodies. Herein, we come across the importance of state-local relations in economic development.[1]

It is needless to emphasize that we have now reached a stage where local finances have merged in state finances and state finances in central finances. Responsibilities have to be matched with resources at all the three levels. Problems of transfer of resources arise at two parallel planes, from the centre to the states and the states to the local bodies. The prospects and pattern of state-engineered economic development depend very much on the manner in which an optimum solution is reached on the question of transfer of resources to match responsibilities.

Conventional discussion on federal finance devotes itself to discovering the principles which should govern the division of responsibilities and the distribution of resources between centre and the states. The discussion then proceeds to examine how equality in sacrifices and benefits could be ensured in a federation. These discussions provide us with valuable insights into questions relating to federal finance but they admittedly recognize that, in practice, federal demarcation does not proceed entirely on rational principles. While principles like independence and responsibility, adequacy and elasticity and administrative economy are important, it is experience and expediency that finally determine the federal demarcation. We shall not therefore cover the well-trodden theoretical ground again[2] but confine ourselves to reflecting on certain broader implications of the distribution of functions and finances.

The inevitable trend of the centralization of taxes and the financial supremacy of the centre is a uniform feature of all federations. By and large variations are in degree rather than substance. It is not necessarily a conscious decision to tip the scales in favour of the centre but the increasing inter-dependence of the federal economy that is contributory to the trends of centralization today. In India, there was no doubt a conscious decision that there should be a strong central government. But the necessities in the matter of centralization of certain taxes have the tendency to make the centre stronger.

[1]State-local relations are treated in some detail in a later chapter. For the importance of local bodies in economic development see the author's article, 'Local Finance in Developing Countries', Journal of Local Administration Overseas, London, July, 1965.

[2]A number of competent studies, Indian and foreign, can be cited. We may refer as examples *Federalism and Economic Growth*, ed. U. K. Hicks (Allen and Unwin), and R. N. Bhargava, *Theory and Working of Union Finance in India*, (Allen and Unwin).

What this obviously means is that the superior resources position of the centre is not because of some superior ability or wisdom on its part in raising resources which the states do not possess, but because of other circumstances beyond the control of the states. This is a fact that central governments can keep reminding themselves of occasionally. Grants-in-aid seen in this perspective do not necessarily have to be instruments of control or direction. They can be conduits of finance, pure and simple, which have to be used where devolution of resources is not possible or expedient. But every purse has its strings and every authority distributing grants-in-aid does not do so without fully enjoying its position as a giver.

The unequal distribution of resources which is such a common feature can be theoretically got over by adjustments in the concurrent list. One can have a concurrent list of taxes as well as functions in a federation. If by convention either the centre or the state should agree to have a particular function performed by one of them rather than by both, the unequal distribution of resources can be successfully got over. For example, we will be raising the question of the possibility of transfer of responsibilities in the sphere of power generation from the states to the centre in order, among others, to reduce the dependence of the states on the centre. But ordinarily while such adjustments in the concurrent lists are theoretically possible there are many practical difficulties. Concurrent taxation in the form of a state surcharge over a federal tax may be useful but it has been found in practice to be economically undesirable. Nor would it be easy to make any central government agree to such a proposal. Similarly, adjustments in the discharge of functions in the concurrent list may be possible in the initial stages of a federation, but it will be difficult at a later stage because surrender of functions is distasteful to any authority.

In practice thus, unequal distribution of resources is not capable of remedy by adjustments in the concurrent list. In so far as this is not possible, the discharge of the various functions of government will vary in standards depending on whether the centre discharges a particular function or the states and among the states which of them discharges it. Grants-in-aid cannot be and have not proved to be a complete corrective to the situation. In so far as they are not a complete corrective, state and local body functions (for this applies to local bodies as well) are doomed to a lower key of performance. Authorities best fitted to perform certain services do not necessarily get the fiscal capacity to perform them best. To the extent that fiscal capacity is injected from outside, the fitness for performance diminishes, if the injecting authority substitutes its judgment for that of the authority competent to judge well.

An obvious example is the disparity in scales of pay of the central and state government staff in India which enables the central government to attract better type of personnel. But the implication of the different standards of performance will be fully understood when it is realized that the states have to perform important functions in education, health, sanitation and road communication.

If the unequal distribution of resources lends scope for 'double standards' in discharge of functions, as between the centre and the states, as a class, it contains at the same time the potentiality for ironing out varying standards as between states. Herein lies one of the great benefits of the unequal distribution of resources. It contains in it the possibility of balanced regional development and the achievement of national minimum standards throughout the federation.

The unequal distribution of resources also means that occasions may arise when certain functions will have to be discharged with much greater efficiency than others and this will be possible only if the centre has sufficient flexibility. For example, if a National Emergency requires diversion of resources for defence purposes, it is much easier to divert resources by constitutional backing than by imploring the states.

The unequal distribution of resources also means that the centre can control and direct the impact of governmental outlays on the economy. It can influence not only the total magnitude but its allocation as well. For example, overall economic considerations in a developing economy may require curbs on expenditure on social amenities like education and health. As a matter of fact, such outlays on social services get rather automatically control since these are functions which the less affluent state governments have to perform.

There is one other important aspect of central control which has not often been realized and which is not strictly a part of the problem of division of tax revenue. In India, the centre's superior financial position and its strong hold over the states is rendered possible not only by its superior power of taxation but by its capacity to borrow. Loans occupy a very important part in the implementation of plan schemes and loan finances have assumed great importance. By controlling the ability to borrow partly by the constitutional provision which requires the consultation of a state with the centre before floating any public loan[1] and partly for other reasons, it is the centre which borrows most of the money and lends to the states for implementation of schemes of a capital nature. It can even be argued that if the states had greater responsibilities for borrowing they might be more careful in borrowing as well as spending. The limitations of borrowing from the open market may be expected to instil in the states a greater sense of financial discipline in capital expenditure. But here again the equalizing role of the centre comes into play as some states like, Orissa, Assam and Kerala will find it difficult to get sufficient capital finance from the open market without the assistance of the centre.

From the perspective of state finances federal financial relations thus form themselves into a kind of triangle where each state looks at the centre as an arbitrary dispenser of grants-in-aid so far as its own finances are concerned and as an arbiter so far as its own finances are compared with that of other states.

It would be interesting to study whether and how the unequal distribution of resources will be altered by economic growth in a developing economy. One view is 'that the most common form of early industrialization in underdeveloped countries involves the use of raw materials that were previously exported to make finished products that were previously imported and this both reduces the revenue from export and import duties increases the internal incomes and profits that are available for regional taxation (provided the regional authorities have the constitutional power to levy such taxes). It follows that though the imbalance of revenues may be a very serious problem in the early years of the federation, it would generally be a mistake to assume that it is more than a temporary problem.'[2] But the experience of no federation, not least that of India, can be said to support this view. Another view is that 'the very process of planned development entails increased revenues to the centre and greater liabilities to the states'[3] since the growth of production in organized industry automatically broadens the resource base of the centre while increasing the need for social services which are the liability of the states. Neither view can be accepted in the form it has been put forward above but it may be

[1]Amplified later.

[2]A. H. Birch in *Federalism and Economic Growth*, ed. U. K. Hicks (Allen and Unwin), p. 116.

[3]Memorandum for the Fourth Finance Commission, Government of Madras, 1965, p. 2.

safe to say, negatively, that planning is not likely to reduce the unequal distribution of resources.

Planning has in any case added fresh dimensions to federal-state relations. Apart from the circumstances in India which will be described in detail in the relevant chapter, a few general aspects of the problem may be noted here. The planning process involves political decisions, economic decisions, financial decisions and administrative decisions. Co-ordinated planning between the centre and the states involves political acceptance of the contents of the plan at both levels and willingness to make the necessary political sacrifices in terms of, say, efforts in additional taxation. The existence of the same political parties in power in both the centre and the states as in India simplifies the political processes in planning which may not be equally simple if different parties held power.[1] Besides, the heavy financial control that the centre may exercise means the substitution of the centre's political or even administrative judgment for the political judgment of the state in purely state subjects. This has both political and constitutional overtones. Politics is also apt to enter the picture in the question of location of central projects.

On the economic side planning in a federation involves economic decisions on the part of the centre for the economy as a whole which in turn can affect the nature and pattern of schemes that the states can undertake. Decisions intended to control the allocation of the national product between consumption and investment, decisions involving choice between projects productive in the long run or short run or decisions favouring the promotion of human capital or physical capital are all ultimately translated into certain patterns of schemes and patterns of expenditure of the central as well as state government. Central assistance will naturally be directed to securing such ends and this may not necessarily coincide with the directions in which the state government would like to incur expenditure.

This apart, there is on the economic side the obvious need for co-ordination and co-operation in tax policy between the centre and the states. This will be necessary not only in the case of direct overlapping between the centre and the states in respect of any particular tax but also in the case of taxes having similar effects or affecting the same tax base as (1) union excise and sales tax and (2) income tax and agricultural income tax in India.

On the financial side planning involves the determination of issues

[1]This has actually happened after the Fourth General Elections, 1967. See the concluding chapter for further discussion of this aspect.

like the nature and quantum of central assistance, the type of grants and the criteria for their eligibility. Central planning will besides involve judgments on the relative backwardness of particular states and decisions on methods for promoting balanced regional development. Administratively, questions will arise impinging on the flexibility of state governments for effective implementation.

Problems like these have arisen in the fifteen years of planning. After all no Constitution can provide an once-for-all solution to federal-state problems. The ensemble of state finances in India is, what one may call, mixed. On the one hand there are complaints that the states as a class and certain states in particular have not pulled their weight in the developmental effort. The easiest criticism is to charge them with improvident spending and indiscreet husbanding of resources but a balanced approach would necessarily have to examine how far the weight is too much for them to pull in the current setup of centre-state financial resources and how far planning itself is adding to their financial stringency. From 1961 onwards some states have had persistent overdrafts with the Reserve Bank and this has been the object of much concern and the subject of frequent homilies from the Union Finance Ministers. By now it has become clear that such overdrafts are not in the nature of temporary or seasonal ways and means advances but are telling symptoms of a deeprooted malady. This has necessarily had the effect of delaying the progress of major projects in certain states.[1] The centre has had to bale them out of these overdrafts by advancing *ad hoc* loans.[2] During the Third Plan the centre assisted eleven states to the tune of Rs. 285·72 crores to clear their overdrafts with the Reserve Bank, of which Rs. 218 crores had been recovered till March 31st, 1967.[3]

[1]'The centre has sanctioned a Rs.7 crore loan to the Andhra Pradesh Government for the Nagarjunasagar project but nevertheless, the state has not found the way of paying off outstanding bills of over Rupees one crore or keeping the work on the project going. Despite repeated requests, the state has no hope of getting funds from the centre in time and the present loan of Rs.7 crores would never reach the state because it would be a 'book adjustment' against the interest the state government has to pay for the past loans' (*The Mail*, Madras, 28th October, 1966).

[2]In 1966–67, the centre advanced loans to seven states totalling Rs. 149·25 crores for clearing their overdrafts with the Reserve Bank. Rough magnitudes: Andhra Pradesh—Rs. 70·00 crores, Assam—Rs. 8·70 crores, Bihar—Rs. 2·00 crores, Madhya Pradesh—Rs. 12·00 crores, Orissa—Rs. 12·00 crores, Rajasthan—Rs.27·95 crores and Mysore Rs. 15·65 crores. Rs. 41·25 crores were recovered from the states in the same year. (*The Mail*, Madras, 14th June, 1967).

[3]*The Mail*, 4th June, 1967.

The other side of the picture shifts part of the blame for this disturbing situation on the centre itself. 'The Planning Commission', according to one view, 'is responsible to some extent for the financial indiscipline in the states. In its passion for big Five Year Plans, it has repeatedly fixed unattainable financial targets. As a result the states have found it difficult to mobilize the required resources'.[1] For nearly ten years now, the field of centre-state relations has never been quiet. The states have raised their voices for more central aid and less strings. The growing and marked dependence of the states on central assistance has been viewed by observers with increasing concern. The phenomenal indebtedness of the states to the centre has brought to the fore problems on the capital as well as the revenue account. The growing financial dependence of the states on the centre in recent years, it is said, is matched only by the marked dependence of the nation as a whole on external assistance.[2]

The pattern of central assistance for plan schemes has been criticized as stifling the initiative and discretion of the state governments and as having degenerated into a set of insipid financial controls without being levers of effective physical implementation. The centre has been accused of taking unilateral financial decisions affecting the interests of the states and of sins of commission and omission in enlarging the divisible pool.[3] Nor have the states been satisfied with the centre in regard to balanced regional development.

Constitutional questions of importance have also arisen. 'No part of the Indian Constitution' says Mr K. Santhanam, the Chairman of the Second Finance Commission and an authority on federal financial relations, 'has been found so inadequate in the context of planning as Part XII which regulates the financial relations between the Union and the States'.[4] As early as 1957 the Second Finance Commission raised the question whether 'the time is not ripe for a review of the constitutional provisions dealing with the financial relations between the Union and the States'.[5] The Third Finance Commission noted the dyarchy in central assistance and suggested that an independent highpower commission should go into the entire question. The Fourth Finance Commission likewise called for a review of Union-State financial relationships. Things have been no

[1]*Capital*, 4th August 1966, p. 241.

[2]I. G. Patel in his foreword to Dibakar Jha, *Bihar Finances*.

[3]See for details in Chapter VI.

[4]*Transition in India* (Asia Publishing House), p. 112.

[5]P. 72, Report of the Finance Commission, 1957, Government of India Press, New Delhi.

better since the Commission reported and Mr Santhanam has now suggested that federal financial relations should be placed on a definite statutory basis by abolishing the institution of Finance Commissions and by making constitutional provisions in respect of allocation of taxes between the Centre and the States.[1]

All this should be enough to convince the reader as well as the author why this book should be written.

[1]A. D. Shroff Memorial Lectures, printed in *Commerce*, 5th and 12th Nov., 1966 and *Indian Finance*, 5th and 12th Nov., 1966.

CHAPTER II

THE BACKGROUND

We shall now briefly set out the factual data necessary for an understanding of our study of the finances of the state governments in the subsequent chapters. A historical account of the growth of state finances and a comparative study of the Indian federation with other federations may not be out of place at this juncture.[1] But we shall avoid this temptation, for any lengthy treatment of these subjects will not be quite germane to the scheme of this book which seeks to provide a perspective analysis rather than a descriptive account of state finances. We shall therefore compress in this chapter a brief historical account as well as what may be called the 'vital statistics' of the *dramatis personae* viz. the states, besides describing the distribution of powers in the Indian Constitution which provides the ultimate framework for the subject of our study.

Before the Montague Chelmsford Reforms of 1919 there was no division of resources between the centre[2] and the provinces. All revenues whether they accrued from land revenue or from customs or from any other source were treated as revenues of the British Crown. From 1912, however, there was a working division of revenues called the system of divided heads of revenue. Yet, the centre had complete control of all taxation in British India and a proposal for provincial taxation during this period required the sanction of the Government of India, the approval of the Secretary of State and the assent of the Finance Department before it could be considered by the Provincial Government. There was similarly strict control and supervision by the central government on provincial expenditure. The provincial governments had also no independent powers of borrowing. The financial system was essentially unitary in character.

[1]The subject of provincial finances in India has been adequately treated in a number of books. See, for example, R. N. Bhargava, *Theory and Working of Union Finance in India* (Allen and Unwin), B. R. Misra, *Indian Federal Finance* (Orient Longmans) etc. For a discussion on the evolution of financial relations, see also Asok Chanda, *Federalism in India* (Allen and Unwin), 1965, ch. IV.

[2]We shall be using the words 'Union' and 'Centre' interchangeably throughout the book.

The Montague Chelmsford Reforms moved towards a kind of division of functions and finances. Subjects were classified into Central and Provincial. To entrust a measure of limited responsible government provincial subjects were classified as reserved and transferred. This system of 'dyarchy' kept subjects like administration of justice, law and order, preservation of financial stability as reserved subjects and only other subjects of a 'nation building' kind were transferred to Ministers who were expected to be responsible to the Legislatures of the Provinces.

Thought there was thus severe financial control over the scope and directions of government expenditure, the distribution of resources strangely enough tended to make the provincial governments richer than the centre and a system of provincial contributions had to be devised to feed the centre. The divided heads of revenue were abolished and land revenue, irrigation, excise, forests and judicial stamps were entirely transferred to the provinces; while customs, other stamps, receipts from railways, salt, opium and Post and Telegraphs were to remain wholly central heads of revenue. Centre-State relations came into prominence from then on and various schemes were proposed to put the financial relations on a satisfactory basis.

The Government of India Act, 1935, restored the financial vantage of the centre, not so much because of a radical redistribution of resources but because the tax potential of the centre was growing. Under this Act, income tax and corporation tax were made exclusively central sources of revenue but agricultural income tax was made a provincial source of revenue. As the provinces were expected to be in deficit, a system of sharing of income tax was contemplated. It was also laid down that salt, excise and export duties could be shared as and when necessary. The devolution of the resources from the centre to the provinces was based on the recommendation of Sir Otto Niemeyer who, among other things, prescribed 50% as the share of the divisible pool of the income tax to be distributed to the provinces.

This was the financial system which India inherited at the time of Independence. The intervention of the Second World War had increased expenditure and receipts substantially but the system remained the same and was adopted in the Indian constitution. But the budgets of the states and the states themselves had to undergo changes and adjust themselves to the new conditions.

Partition affected the finances of the states of Bengal and Punjab. The merger of the princely states with the erstwhile British India had its impact on the finances of many states. The administration of these states had to be brought on a par with that of the erstwhile provinces of the Government of India. Varying standards of services had to be brought to a uniform basis. There was a certain amount of financial benefit (though it does not seem to have been fully assessed) by the extension of taxes to the areas of the erstwhile princely states but the expenditure to be incurred was even more. According to the data of the census of 1941, the population of the nine British provinces (allowing for partition) was 228 million while the population of the Indian states merged with these provinces was 19·5 million. The addition to population was the largest for Bombay (85 lakhs), Orissa (51 lakhs) and Madhya Pradesh (28 lakhs).[1] States like Rajasthan which were formed entirely out of princely states had of course these problems in a greater measure. The abolition of customs duties involved financial loss and the enlargement of the police force due to the disbandment of the army increased the commitments. The financial effects of the mergers were no doubt taken care of by several measures but there is no doubt that the mergers entailed financial and administrative problems for the states.[2]

Independent India had therefore two problems to reckon with in regard to economic development. One was the disparities between the erstwhile British provinces and the erstwhile Independent states and the other the disparities among the provinces themselves. Surprisingly enough the per capita levels of expenditure have not fully reflected the disparities in economic development. Table 2.1 will show the per capita revenue of the various states in 1950–51, that is just before the First Five Year Plan. It is an interesting table even though it is for an individual year and no allowance is made for abnormal expenditures.

It will be seen that Part B states[3] were not so badly off in respect of their levels of expenditure as might be imagined. Their tax revenues per capita were themselves higher than that of the Part A states. It will however be seen that there were wide variations in the levels of expenditure among the states, including the Part A states.

[1] *Reserve Bank of India Bulletin*, 1951.

[2] The question is also referred to in the next chapter. As regards the financial problems, see, for example, the Report of the Rajasthan Finance Enquiry Committee, 1958.

[3] The First Schedule to the Constitution distinguished three types of territories listed as Part A, Part B and Part C respectively. Part A comprised the erstwhile British provinces and the princely states merged in them. Part B consisted of new territories creased out of princely states and Part C states consisted of territories administered directly by the Union. The distinction was abolished in 1956 after the reorganisation of states.

Bombay had the highest per capita revenue and standards of security and social services followed by states like West Bengal, Punjab and Madras. On the other hand, states like Bihar, Orissa, Madhya Pradesh and Uttar Pradesh had poor standards of public revenue and expenditure. One of the important problems of a federation is to reduce the disparities in levels of governmental services among various states and this problem confronted the planners from the beginning.

TABLE 2.1

Per Capita Revenue and Expenditure of States in 1950–51

(actuals, rupees)

				Expenditure on	
Part A States	Revenue	Tax Revenue	Expenditure	Security Services	Social Services
Assam	10·59	7·59	9·98	2·00	3·05
Bihar	7·14	5·50	6·42	1·99	1·81
Bombay	17·52	13·51	17·54	5·77	5·90
Madhya Pradesh	9·14	7·67	7·79	2·27	2·15
Madras	10·06	7·97	10·29	3·10	3·51
Orissa	7·12	5·12	8·17	2·29	2·44
Punjab	13·28	8·15	12·60	3·98	3·26
Uttar Pradesh	8·11	6·07	8·10	2·82	2·54
West Bengal	13·67	11·39	14·87	3·84	3·89
All Part A States	10·42	8·03	10·38	3·19	3·21
Part B States					
Hyderabad	13·85	10·59	14·58	4·51	3·23
Madhya Bharat	12·98	9·24	14·71	4·24	4·41
Mysore	15·48	6·97	14·53	1·99	7·03
Pepsu	17·26	11·29	14·37	4·11	3·40
Rajasthan	9·43	7·46	8·97	3·05	2·50
Saurashtra	18·50	8·60	17·67	6·62	5·52
Travancore-Cochin	14·73	9·60	13·41	1·75	4·16
All Part B States	13·55	9·02	13·34	3·54	3·99

Note: 1. 'Security services' include general administration, administration of justice, jails and convict settlements, police, ports and pilotage and miscellaneous departments.

2. 'Social services' include scientific departments, education, medical, public health, agriculture, rural development, veterinary, co-operation, industries and civil aviation.

[Source: Reserve Bank of India Bulletin, Nov. 1952, p. 913.]

The reorganization of states added to this problem and had its own impact, favourable as well as adverse, on state finances and development. The major changes of reorganization apart from the integration of the princely states were (1) the partition of Madras State and the creation of Andhra State in 1953; (2) the reorganization of states on a linguistic basis in November 1956; and (3) the partition of Bombay state into Maharashtra and Gujarat in 1960. It is not necessary for us to go into the details of the reorganization of the states here.[1]

Even after reorganization, there are still wide disparities in the size and population of various states and this clearly has implications for the economic development of the various states. The historical tradition of a state, whether it was largely a part of erstwhile British territory or whether it was largely a part of the erstwhile Indian States, has much to do with the standards of administration. Other factors like the size of a state, its natural resources etc. will also play a part in determining the kind of development that a state is capable of.

It is difficult to compress in a short space a comparative study of the basic indices of development in various states. We have no alternative but to inflict four tables (Tables 2.2 to 2.5) on the reader to give an idea of the kind of entities the finances of which we are dealing with. We have chosen the year 1960–61 as that is the year for which figures, particularly those of state income, are readily available. We are thus not making a comparison over time but of states after nearly ten years of planning. Here we inevitably come up against questions of the relative levels of development of various states which will in a sense require to be taken up later. But there is an advantage in getting to know inter-state disparities at the outset so that the mention of any particular state in discussion evokes associations of the relevant type. The tables contain a range of materials which can be annotated at length. But we shall leave this to the calm reflection of the reader and confine ourselves to spotlighting some important and relevant aspects.

TABLE 2.2 enumerates the area, population, income and per capita income of the states and expresses their tax revenues as a percentage of state income. The share of Andhra Pradesh, Bihar, Jammu and Kashmir, Kerala. Madhya Pradesh, Mysore, Orissa, Rajasthan and Uttar Pradesh in national income is less than proportionate to

[1]From October, 1966, instead of the State of Punjab, we have two states, Punjab and Haryana. As already stated, we shall not be taking note of this change in this work.

their share in the national population. Assam and Madras have an equiproportionate share while Gujarat, Maharashtra, Punjab and West Bengal have a more than proportionate share. The tax revenue as a percentage of state income also varied from state to state, though the variation has not been very significant except for the low percentages of 1·9 in Orissa and 2·3 in Gujarat. For others, the percentages vary between 2·8 to 3·9.

TABLE 2.2

Area, Population and Income of States

States	Area		Population	
		Percentage		
	Sq. miles (000)	Distri- bution	Number (Lakhs)	Percentage
1. Andhra Pradesh	106	8·8	356	8·2
2. Assam	47	3·9	119	2·7
3. Bihar	67	5·6	465	10·6
4. Gujarat	72	6·0	206	4·7
5. Jammu and Kashmir	55	4·6	36	0·8
6. Kerala	15	1·2	169	3·9
7. Madhya Pradesh	171	14·2	324	7·4
8. Madras	50	4·2	337	7·7
9. Maharashtra	119	9·9	396	9·0
10. Mysore	74	6·2	236	5·4
11. Orissa	60	5·0	175	4·0
12. Punjab	47	3·9	203	4·6
13. Rajasthan	132	11·0	202	4·6
14. Uttar Pradesh	114	9·4	737	16·8
15. West Bengal	34	2·8	349	8·0
All India including other territories	1,203	100·0	4,381	100·0

Note: Population is according to 1961 census. State income is as in 1960–61.

[Source: Pp. 6 and 9, *Distribution of National Income by States*, 1960–61, NCAER and others.]

TABLE 2.3 depicts the agricultural and power development in the states. It should be remembered that the size of a state and its physical features determine the upper limit of such development. But it is not incorrect to assume that the upper limit has not been reached in any state. The figures relating to production of food grains clearly indicate how some states have a job of feeding their population from the surpluses of other states. Agricultural production therefore becomes a national problem and the centre has to give a higher priority to agricultural production in all states (including surplus states), than the surplus states themselves might.[1] The figures relating to gross irrigated area provide an index of how much the dependence on the monsoon is partial or total in each state. Power development has been uneven in the initial years, with the progressive states and those with multipurpose schemes forging ahead. The

TABLE 2.2—*Continued*

Net Domestic Product		Per Capita Income	Tax Revenue as percentage
Rs. Crores	Percentage Distribution	Rs.	of state income
1,033	7·1	287·9	3·6
396	2·7	333·3	3·2
1,025	7·0	220·7	3·2
812	5·5	393·4	2·3
103	0·7	289·0	..
532	3·6	314·9	3·1
924	6·3	285·4	2·8
1,125	7·7	334·1	3·2
1,853	12·6	468·5	3·8
719	4·9	304·7	3·9
485	3·3	276·2	1·9
917	6·3	451·3	2·8
539	3·7	267·4	3·1
2,193	15·0	297·4	2·7
1,623	11·0	464·6	2·8
14,653	100·0	334·5	..

responsibilities of the state governments for irrigation and power are immense indeed.

From Tables 2.4 and 2.5 we have an idea of the diversities in social and industrial development. As we have stressed already our idea in presenting these tables is not for a regular discussion on inter-state imbalances but for stressing the inevitable existence of diversities and for familiarizing the reader with the *dramatis personae* which will otherwise be mere names and units.

[1]This is said to be a reason for central control on plan aid. See the minute of dissent of Mr. G. R. Kamat to the Third Finance Commission, 1961.

TABLE 2.3

Agricultural and Power Development in States

State	Percentage in total population	Percentage in food grains production	Percentage of net irrigated area to net sown area	Electricity generated per capita (kwh)
(1)	(2)	(3)	(4)	(5)
Andhra Pradesh	8·2	8·5	27·1	29·9
Assam	2·7	2·1	28·1	3·6
Bihar	10·6	9·2	22·6	32·5
Gujarat	4·7	2·8	6·7	65·7
Jammu and Kashmir	0·8	0·8	45·1	20·2
Kerala	3·9	1·4	18·7	40·7
Madhya Pradesh	7·4	11·9	5·8	20-9
Madras	7·7	6·8	38·6	74·8
Maharashtra	9·0	8·3	5·7	91·3
Mysore	5·4	5·0	7·9	46·4
Orissa	4·0	4·8	17·4	33·9
Punjab	4·6	7·9	40·1	91·0
Rajasthan	4·6	6·4	10·9	7·0
Uttar Pradesh	16·8	17·7	30·1	18·4
West Bengal	8·0	6·3	25·8	75·7
All States	100·0	100·0	18·1	44·6

Note: Col. (2)—The percentage is for all India including other territories. Col. (3)—percentage includes that of Nagaland. Figures relate to average of 1958–59 to 1962–63. Col. (4)—for the year 1959–60. Col. (5)—for the year 1961–62.

[Source: Tables 5 and 7 in Appendix to *Report of Finance Commission*, 1965.]

It is not possible to devote more time and space to these aspects at this juncture but we must naturally ask what policy implications they have for the financial expert and the planner. Obviously the problem of inter-state disparities cannot be approached in a mechanistic way as one of levelling or making up some percentages.[1]

[1]The problem of inter-state disparities is touched upon in the concluding chapter also. For a descriptive account of questions relating to regional balance in India, see A. H. Hanson, *The Process of Planning*, ch. IX. (Oxford University Press, 1966). For a theoretical and mathematical approach to equalization and the problems involved in it, see R. A. Musgrave, "Approaches to a Fiscal Theory of Political Federalism" in *Public Finances, Needs, Sources and Utilization*, A Report of the National Bureau of Economic Research (Princeton, 1961) and K. V. S. Sastri, *Federal-State Fiscal Relations in India* (Oxford University Press, 1966).

People's sense of priorities in different regions varies. If each state were given Rs. 5 crores to spend as it liked, all of them would not spend the money in an identical way. The centre's sense of priorities may be different from that of particular states. Thus regional imbalances are partly the result of state policies also. Bringing up the levels on a par with those of other states also involves considerable time and expenditure which the state finances will not be able to bear on their own all of a sudden.

The entire question in turn revolves on the distribution of resources and functions. We may now study the pattern of distribution of functions and resources between the centre and the states.[1] The

TABLE 2.4

Roads, Education and Hospital Beds in States (1960–61)

State	Road mileage * per 100 sq. m.	per 1000 population	Education % of age group 6-11 at school	Beds per million population
Andhra Pradesh	14·63	0·45	60·3	486
Assam	1·96	0·15	77·4	337
Bihar	10·62	0·16	53·5	219
Gujarat	..	..	72·0	291
Jammu and Kashmir	0·94	0·25	45·0	791
Madhya Pradesh	6·73	0·39	47·0	377
Madras	33·37	0·52	78·9	667
Maharashtra	10·71	0·37	73·3	253
Orissa	6·32	0·23	47·8	228
Punjab	11·51	0·29	61·8	724
Rajasthan	5·66	0·41	42·0	323
Uttar Pradesh	10·61	0·17	45·4	240
West Bengal	18·11	0·20	65·6	743
All States	10·87	0·33	60·7	445

* 1959 March

[Source: Pp. 24–25, *Economic Development in Different Regions in India*, Planning Commission, 1962.]

[1]At the time of Constitution-making, while the provinces were keen on a larger share-out, they do not seem to have been particularly exercised over questions of 'state autonomy' and their implications for the distribution of resources. Of course, they could not have foreseen the difference that planning would make. On the framing of the financial provisions of the Constitution, see Granville Austin, *The Indian Constitution, Cornerstone of a Nation* (Clarendon Press, 1966), ch. IX. On the question of division of powers and allied issues, see also A. Krishnaswami, *The Indian Union and the States—A Study in autonomy and integration* (Pergamon Press, 1965).

TABLE 2.5

Social and Industrial Conditions in States (1960–61)

State	Density per square mile	Percentage of urban population	Percentage of scheduled castes and tribes *	Per capita value added by manu-facture (Rs.)
Andhra Pradesh	339	17·45	17·5	7·68
Assam	252	7·69	23·6	22·97
Bihar	691	8·43	23·1	14·46
Gujarat	286	25·76	20·0	48·82
Jammu and Kashmir	..	16·66	7·5	2·89
Kerala	1,127	15·11	9·6	14·11
Madhya Pradesh	189	14·29	33·8	8·80
Madras	669	26·69	18·8	23·93
Maharashtra	333	28·22	11·7	67·35
Mysore	318	22·33	14·0	14·47
Orissa	292	6·32	39·8	6·87
Punjab	430	20·13	20·5	11·93
Rajasthan	153	16·28	28·1	5·22
Uttar Pradesh	649	12·85	20·9	8·44
West Bengal	1,032	24·45	25·8	58·24
All States	383	17·64	21·4	22·47

*Such castes, races or tribes as are deemed to be so under Articles 341 and 342 respectively of the Constitution. They cover the most backward and underprivileged groups in the country.

[Source: *Report of the Finance Commission*, 1965.]

division of powers and resources is listed in the 7th Schedule of the Constitution which is reproduced *in toto* as an Appendix to this book for the sake of easy reference. We shall confine ourselves here to the enumeration of the main points of note in this system of distribution. There are three lists, the Union List, the State List and the Concurrent List. The Concurrent List enumerates a number of functions but does not include any power of taxation—which clearly excludes problems of federal-state tangles in taxation.

Before we refer to individual items of functions and finances, it is necessary to point out that the scheme of the Constitution provides for the superiority of the Centre over the states in certain respects. Legislation by Parliament on any subject in the Union List prevails over any legislation by a State Legislature on any subject included in the other lists. Any legislation of the Union on a subject in the Concurrent List will be valid as against any legislation passed by a State Legislature on the subject. More importantly, by Article 249, the Council of States is empowered by a two-thirds majority to transfer to the Union or Concurrent List any subject in the State List for one year at a time. Article 250 empowers Parliament to legislate on subjects in the State List, when a state of national emergency is declared by the President. Article 353 enables executive decisions to be given by the Union Government to the State Government in respect of matters necessary to meet a national emergency. Article 356 empowers the President to suspend the State Constitution and transfer to himself all the powers of a state government and Legislature, when in the opinion of the Governor of the state or the Union Government, the government of that state cannot be carried on in accordance with the Constitution.

To revert to the division of powers, the centre has responsibilities for defence, external affairs, foreign trade, railways, currency and coinage, posts and telegraphs, regulation of inter-state commerce, regulation and development of inter-state rivers and river valleys, insurance, atomic energy, elections, audit and accounts etc. Functions relating to law and order, police, jails, administration of justice etc. are with the States. Subjects like economic and social planning, electricity, public health, education, land, irrigation, agriculture, social security, price control etc. are in the concurrent list.

As regards the distribution of resources, there are as many as nineteen items of revenue in the state list such as land revenue, excise on liquor and opium excluding medicinal and toilet preparations, stamps other than those in the Union List, agricultural income tax, sales and purchase tax (other than on newspapers), tax on lands and buildings, terminal tax on passenger and goods, tax on sale of electricity, tax on vehicles, animals and boats, amusements, betting and gambling, and professions and callings. These taxes can be levied either by the state government or they may be transferred by state legislation for exploitation by local bodies.

The twelve items in the Union List fall into five groups. (1) Exclusively Central: Under this fall customs including export duties, corporation tax, taxes on the capital value of assets of individuals and companies exclusive of agricultural land. A surcharge on income tax is also exclusively central by virtue of Article 271 of the Constitution. (2) Taxes on income other than agriculture income: The centre levies and collects the income tax but it is bound to share the income tax with the states, on a basis to be recommended by the Finance Commission. (3) Union Excise duties (other than on liquor and opium) are levied and collected by the centre but may be shared with the states if Parliament so desires. (4) Taxes which are levied and collected by the centre but to be entirely distributed to states by the Parliament. Succession and estate duties, terminal tax on passengers and goods carried by air or sea, tax on railway fares and freights, tax on sale or purchase of newspapers; and sale or purchase tax on inter-state trade. (5) Taxes levied by the centre but to be collected and appropriated by the states: Stamp duties on bills of exchange, cheques, promissory notes etc. and excise duties on medicinal and toilet preparations containing alcohol.

The Constitution has thus recognized the need for the transfer of resources from the centre to the states and provided for it. It has also recognized that it would be necessary to give grants-in-aid to the states. There are two Articles under which grants-in-aid could be made: (1) Under Article 275 grants-in-aid may be given to such states as Parliament may determine to be in need of assistance; different amounts may be fixed for different states. The grants-in-aid under this Article are to be determined with reference to the recommendations of the Finance Commission. (2) The other Article which provides for grants-in-aid is Article 282. Though it has been used widely for financing the Five Year Plans, it actually comes under the category of 'Miscellaneous financial provisions'. Article 282 provides that the Union or the States may make any grants for any public purpose notwithstanding that the purpose is not one with respect to which Parliament or Legislature of the State as the case may be, may make laws.

The borrowing powers of states are particularly important in the context of planning. Under Article 293, the Government of India can give loans to states. The states can also borrow from the open market within India but the consent of the Government of India is necessary so long as there is an outstanding loan of the centre to the state concerned. This has virtually meant that the consent of the Government of India and through it, that of the Reserve Bank, is always necessary, since central loans to states not only exist but are growing in a phenomenal measure. Under the Article the consent is necessary only for the act of borrowing and not its quantum but this is in practice a distinction without a difference.

The actual machinery for transfer of resources from the centre to states is through the Finance Commission which is provided for in the Constitution and the Planning Commission which is not. The Finance Commission is a unique feature of the Indian Union. It is a five-member Commission appointed once in five years by the President for making recommendations on (1) distribution of the net proceeds of taxes which are to be or may be divided; (2) the principles governing the grants-in-aid from the centre to the states; and (3) other matters which may be referred to the Commission by the President in the interest of sound finance.

The Finance Commission does not deal with loans, unless of course the President refers the question to it. Even in respect of grants, the Finance Commission's pastures have been restricted by circumstances to the non-Plan side. The quantum of loans and the plan grants to states is determined by the Planning Commission with reference to the size of the plan of the state, its revenue and capital resources, its needs and so on. This dualism in the source of financial system has raised 'first-class' problems in centre-state relations which will be dealt with in later chapters.

CHAPTER III

OVERALL TRENDS

We shall now attempt to study the overall trends in the finances of state governments from Independence and particularly from the beginning of the First Five Year Plan in 1951. Before we turn to the relevant figures it is necessary (1) to underline the difficulties in comparison and (2) to describe briefly the structure of the Budget of a state government. We shall do this in Part I of this chapter before getting to the overall trends in Part II.

PART I

The difficulties in comparison are of a substantial nature and may be divided into two categories; (1) those arising from territorial changes due to re-organization and (2) those arising from changes in presentation and classification. But these should not discourage us from a comparative study since such a comparative study is quite necessary for our gaining some useful knowledge of state finances. What is necessary is only that we should remember the limitations of the comparisons.

After the beginning of the First Five Year Plan major reorganizational changes arose from (1) the creation of Andhra Pradesh (2) States' Re-organization in 1956 and (3) creation of Gujarat State. We have already indicated in the previous chapter the changes implied by all these re-organizations. Because of such changes, comparisons over time of the figures of the same state as well as comparisons among states *inter se* are both difficult. The figures for 1956–57 are available only in a truncated form in terms of pre-reorganization and post-reorganization periods. Broadly speaking, comparison for the First Five Year Plan is possible for the Part 'A' and Part 'B' states except for the creation of Andhra State. Comparison for the Second Five Year Plan period is possible except for the year 1956–57 and the bifurcation of Bombay State. Comparison is fully possible for the Third Five Year Plan.

The other set of difficulties arises from the changes in presentation and classification. For one thing, it should be remembered that the figures here which are based on the figures presented by the Reserve Bank of India Bulletin will be different from the figures in the state budgets, since the object of the Bulletin is to show the real budget of the state rather than an accountants' budget. From 1960–61, the Bulletin has exhibited revenue receipts net of transfers from and to reserve and similar funds. Net sales of securities have also been excluded from capital receipts. Subject to these changes, however, the Bulletin has followed, by and large, the same procedure from the beginning. Other important presentational changes are enumerated below. (1) Commencing from 1956–57, works of a capital nature the cost of which exceeds Rs. 20,000 as also similar works which form part of a scheme whose total cost exceeds Rs. 1 lakh are presented as capital expenditure outside the Revenue Account. Until then though they satisfied the definition of capital in the Audit Code, they were being shown under revenue expenditure as the monetary limit for inclusion in capital was much higher. In the states' budgets certain receipts such as those from the Central Road Fund which were formerly credited to revenue with the corresponding expenditures being debited to expenditure were now shown as reduction of expenditure, thereby reducing both revenue and expenditure totals. Grants to local bodies and other public institutions for works (material assets) could also now be debited to capital, provided the expenditure was written back to revenue over a period of, say, fifteen years. The provision for repayment of both market loans and borrowings from the centre was now to be shown only in the public debt section. The net result of these changes was a fairly substantial relief to the revenue budget. The overall budgetary position including the Public Account will however remain the same. (2) The constitution of the State Electricity Boards has affected the capital budgets of the state. Electricity undertakings hitherto managed departmentally were brought under statutory electricity boards created by most states during the Second Plan. When electricity undertakings were managed departmentally, the capital expenditure was shown as capital expenditure outside the revenue account. After the constitution of the electricity boards the capital expenditure was undertaken in most states by electricity boards which were given loans by the respective governments.[1] The entire transactions of the electricity board are conducted in the 'personal deposit account' kept by it in the Public Account of the government.

[1]For a detailed description of the position, see Chapter XII.

The overall budgetary position is again not affected by this change. (3) From 1962–63 the recoveries of interest on the capital advanced to commercial departments like Irrigation, Electricity etc. were shown as gross receipts as against the procedure of adjusting the accounts in reduction of interest charges followed up to 1961–62. This has inflated the revenue as well as the expenditure of the state governments in the revenue account. The surplus or deficit in the revenue account and in the overall budgetary position would however remain the same.

We will therefore have to be careful in the comparison of figures over years. We shall ourselves present the figures for each Five Year Plan period separately in this chapter so that comparisons are not vitiated.

It is now necessary to give a brief idea of the structure of the budget of the state governments. The budget of a state government is drawn up on the basis of actual cash receipts and expenditure[1] for each year beginning from the 1st April. It is divided into three parts.[2] (I) The Consolidated Fund, expenditure out of which should be either charged or voted by the legislature; (II) Contingency Fund to meet expenditure on schemes, funds for which have not been voted by the legislature, till such time as the funds are voted; and (III) the Public Account which includes all other transactions of the government of various kinds. The Consolidated Fund in turn has three divisions. The first division deals with the proceeds of taxation and other receipts classed as revenue and the expenditure met therefrom, the net result of which represents the revenue surplus or deficit for the year. The second division deals with expenditure met usually from borrowed funds with the object either of increasing concrete assets of a material character or of reducing recurring liabilities such as those for future pensions by payment of the capitalized value. It also includes receipts of a capital nature intended to be applied as a set-off to capital expenditure. The third division comprises loans raised by government, loans of a purely temporary nature classed as floating debt (such as Treasury Bills and Ways and Means Advances) as well as other loans classed as 'Permanent Debt'

[1]For commercial undertakings accounts in commercial form are also maintained.

[2]Prior to the coming into force of the Constitution, there were four divisions viz. Revenue, Capital, Debt and Remittances. See p. (vi), Combined Finance and Revenue Accounts of the Central and State Governments, 1960–61, Comptroller and Auditor-General of India. Our description of the budget follows this publication.

and 'Loans and advances made by Government' together with payments of the former and recoveries of the latter.

Part III, the Public Account has two divisions, (1) Debt and Deposits and (2) Remittances. The first division comprises receipts and payments other than those falling under debt heads pertaining to Part I in respect of which government incurs a liability to repay the moneys received or has a claim to recover the amounts paid together with repayments of the former and recoveries of the latter. Transactions of various funds like the sinking fund, depreciation reserve funds, local bodies and quasi-government bodies and the state provident fund balances, all figure in this division. The second division embraces merely adjusting heads under which appear remittances of cash between treasuries, transfers between different accounting circles and remittances between India and England. Credits and debits taken to the adjusting heads in this division are eventually cleared by adjustment under final heads.

In Parts I and III transactions are grouped into sections which are further subdivided into Major Heads of Account. The sections are distinguished by letters of the English alphabet, a double letter showing the capital portion of the particular set of transactions. The Major Heads in revenue and capital transactions are numbered serially, Roman numerals being employed on the receipts side and international numerals on the disbursement side.

The combined effect of the transactions in the Consolidated Fund, the Contingency Fund and the Public Account produces the overall budgetary position and the surplus or deficit thereof. The opening cash balance (and securities) of the government account added or subtracted by the surplus or deficit in the overall transactions of the year produces the closing cash balance (and securities) for the year.

The principle that the revenue budget should always be balanced has been given up though it is expected that over a period of years a balance should be achieved. A surplus in revenue account will, if other transactions remain the same, increase the cash balance while a deficit will reduce the cash balance. A deficit in revenue account will therefore be financed by a drawal on the cash balance. If other transactions do not balance, as may usually be the case, a surplus or deficit in the revenue account will improve or worsen the net position under the other transactions. Overall budgetary surplus will increase the cash balance and overall deficit will decrease the cash balance. Apart from this, each state government may have invested some amounts in securities and a

3.0 DIAGRAM OF THE ANATOMY OF THE STATE BUDGET

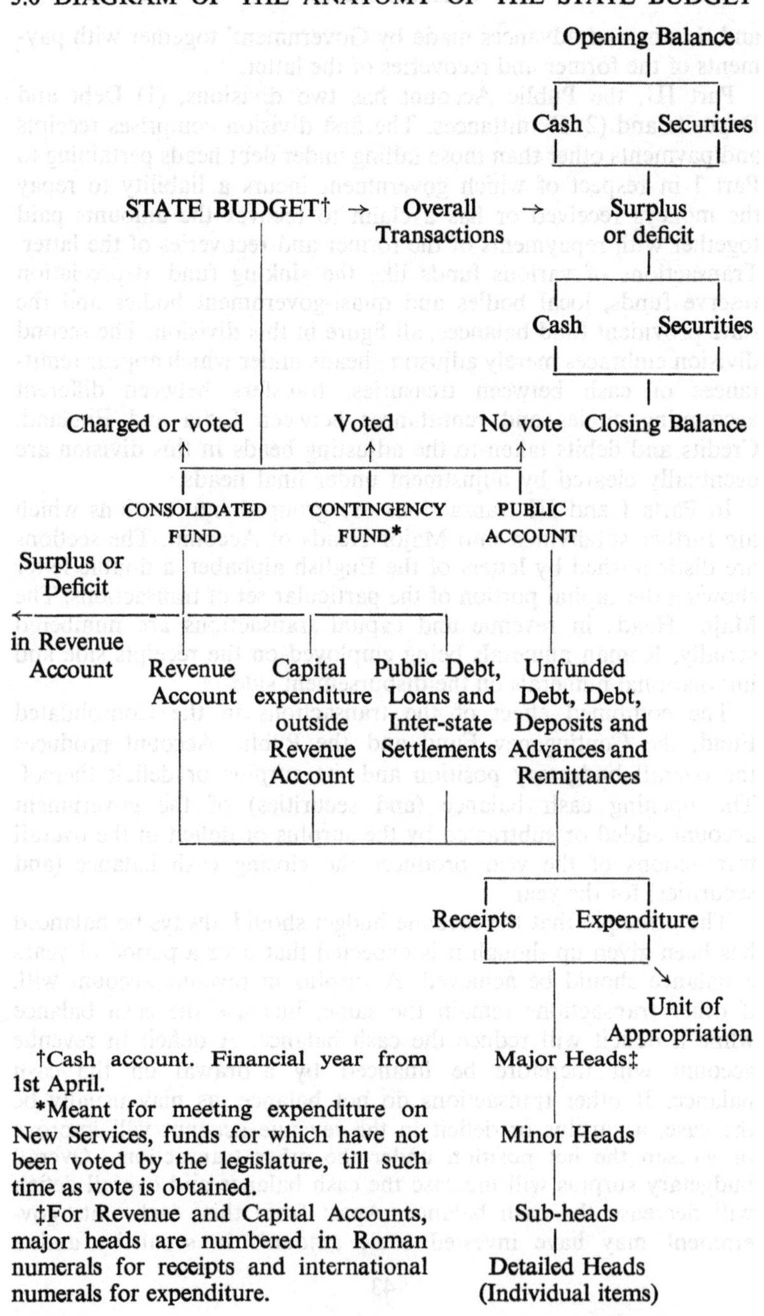

†Cash account. Financial year from 1st April.

*Meant for meeting expenditure on New Services, funds for which have not been voted by the legislature, till such time as vote is obtained.

‡For Revenue and Capital Accounts, major heads are numbered in Roman numerals for receipts and international numerals for expenditure.

surplus in the overall budgetary position may be utilized for increasing the reserve of securities. Similarly, a deficit in the overall budgetary position may be met by a discharge of securities on hand.

While the accounts for each year will reflect the overall transactions of the year, the cash balance of the state government will vary from day-to-day. Receipts and expenditure do not match in time from month to month and the daily variations in the cash balance are known as the 'ways and means' position of the state government. Each state government keeps an account of its cash balance in the Central Accounts Section of the Reserve Bank of India which communicates the cash balance position to the state governments every day. It also invests the cash balance or discharges the securities according to the instructions of the state governments. With reference to the securities that a state government has and other considerations, the Reserve Bank of India, by agreement with the state government, fixes an authorized overdraft limit for each state. Any overdraft over and above the authorized level is called an unauthorized overdraft.

The above in brief is the anatomy of a state government budget.[1] (See Diagram 3.0) There are other aspects of the question of exhibition of state finances, like the formulation of the plans of the states, classification of expenditure etc. and they will be referred to later at appropriate places.

PART II

At the beginning of the First Five Year Plan there was no particular imbalance in the finances of the state governments. Table 3.1 will show the revenue and expenditure on revenue account of the various Part 'A' states before and after the war and at the beginning of the First Plan.

The war increased the revenue and expenditure very substantially, by roughly 250%. In most states the increase was of the order of

[1]The distinction between the actual cash balance on the 31st March, the last day of the financial year and the closing balance of the year should be noted. Even after 31st March, the accounts are kept open for about two months for book adjustments. After this, the accounts closing balance is struck and it is the true closing balance of the year which is presented in budget documents.

TABLE 3.1

Revenue Accounts of States before Planning*

(Rs. Crores)

State	1938–39 R	1938–39 E	1938–39 ±	1947–48 R	1947–48 E	1947–48 ±	1948–49 R	1948–49 E	1948–49 ±	1949–50 R	1949–50 E	1949–50 ±	1950–51 R	1950–51 E	1950–51 ±
1. Assam	2·58	2·99	−0·41	6·62	6·85	−0·23	9·24	9·56	−0·32	10·30	9·94	−0·36	9·92	9·28	0·64
2. Bihar	5·24	4·93	0·31	18·50	16·80	1·70	23·10	21·76	1·34	25·95	25·80	0·15	28·97	26·05	2·92
3. Bombay	12·45	12·80	−0·35	43·73	41·33	2·40	49·91	47·06	2·85	61·53	61·52	0·01	64·31	64·37	−0·06
4. Madhya Pradesh	4·27	4·71	−0·44	12·25	11·36	0·89	17·38	16·60	0·78	19·60	19·27	0·04	19·65	16·74	2·91
5. Madras	16·13	16·10	0·03	56·71	50·69	0·02	53·33	53·33	—	55·89	55·54	0·35	58·16	59·45	−1·29
6. Orissa	1·82	1·81	0·01	6·04	5·84	0·20	6·20	7·34	−1·15	10·82	11·47	−0·65	10·31	12·01	−1·70
7. Punjab	11·36†	11·61†	−0·25	7·45	9·31	−1·86	18·08	16·89	−1·19	16·95	16·12	0·83	16·86	16·00	−0·86
8. Uttar Pradesh	12·80	12·80	—	38·74	38·73	0·01	49·20	49·18	−0·02	56·26	56·26	—	51·89	51·84	0·05
9. West Bengal	12·77†	12·77†	—	18·73	13·28	5·45	31·77	29·10	2·67	34·01	31·38	−2·63	34·30	37·33	−3·03
TOTAL	79·42	80·52	−1·10	202·77	194·19	8·58	258·21	250·82	7·39	291·31	287·29	4·02	294·37	293·08	1·29

*Only 'Part A States'. †Pre-partition figure.

R = Revenue E = Expenditure

[Source: Reserve Bank of India Bulletins.]

300%. Land revenue was still prominent in many states while sales tax began to assume special importance. Between 1938–39 and 1950–51 the receipts and expenditure went up by nearly 300%. This happened in spite of the policy of prohibition which considerably reduced the receipts in certain states. Table 3.2 will show the receipts under 'Provincial Excise' before and after the war and after prohibition. Bombay and Madras lost as much as Rs. 16 crores and Rs. 4 crores respectively between 1945–46 and 1949–50. In spite of this, their receipts continued to be buoyant.

We have already seen the relative levels of per capita revenue and expenditure in various states as in 1950–51. By that time the financial position of the states was fairly balanced and no state had a surplus or deficit over Rs. 3 crores. From the beginning of the First Five Year Plan there was a marked deterioration in the consolidated financial position of the states. Revenue surpluses gave way to deficits amounting to Rs. 1·7 crores in 1950–51, 6·59 crores in 1951–52 and 29·46 crores in 1952–53. Side by side with revenue deficits there were heavy deficits on capital account. This resulted in overall deficits resulting in the running down of cash balances and withdrawal of reserves.

The unsatisfactory overall position was the result of a number of adverse factors such as (1) the effect of partition in some states; (2) political and financial integration involving new commitments;

TABLE 3.2

Receipts from Provincial Excise

(Rs. crores)

State	1938–39	1945–46	1949–50 RE
Madras	3·72	16·80	0·46
Uttar Pradesh	1·33	5·95	5·76
Bombay	2·90	8·91	4·97
Bihar	1·20	4·62	4·90
Madhya Pradesh	0·64	2·44	2·16
Assam	0·35	0·82	0·68
Orissa	0·33	0·88	1·65
West Bengal	1·59*	8·17*	6·16
Punjab	1·02*	3·03*	2·21
TOTAL	13·08	51·62	28·95

* Pre-partition figures.

[Source: Reserve Bank of India Bulletin, 1950.]

TABLE 3.3

Revenue and Expenditure of States in First Plan

(Rs. Lakhs) (actuals)

Part A States	1951–52 R	1951–52 E	1951–52 ±	1952–53 R	1952–53 E	1952–53 ±	1953–54 R	1953–54 E	1953–54 ±	1954–55 R	1954–55 E	1954–55 ±	1955–56 R	1955–56 E	1955–56 ±
1. Andhra Pradesh	..	..	..	..	..	..	..	..	..	1,931	2,449	– 518	2,355	2,884	– 529
2. Assam	1,081	1,045	36	1,385	1,215	170	1,390	1,331	59	1,571	1,869	– 298	2,204	2,445	– 242†
3. Bihar	3,379	3,231	148	3,570	2,773	797	3,317	3,110	207	3,535	4,148	– 613	4,106	5,445	– 1,339
4. Bombay	6,152	6,140	12	6,118	6,712	– 594	7,103	6,988	115	7,801	7,008	793	8,437	7,817	620
5. Madras	5,809	6,310	– 501	5,705	6,703	– 998	6,438	6,438	—*	4,235	4,627	– 392	5,210	5,456	– 246
6. Madhya Pradesh	2,242	1,704	538	2,295	1,830	465	2,402	2,377	25	2,722	2,776	– 54	3,072	3,063	9
7. Orissa	1,164	1,054	110	1,224	1,114	110	1,189	1,283	– 94	1,356	1,505	– 149	1,611	2,319	– 708
8. Punjab	1,709	1,537	172	1,755	1,571	184	1,976	1,890	86	2,183	1,955	228	2,503	2,753	– 250
9. Uttar Pradesh	5,139	5,133	6	6,072	6,072	—	7,139	6,856	283	7,282	7,221	61	8,553	8,423	130
10. West Bengal	3,805	3,677	128	3,691	3,839	– 148	3,746	4,435	– 689	4,196	4,846	– 650	5,033	6,106	– 1,073
TOTAL	30,480	29,831	649	31,815	31,829	– 14	34,700	34,708	– 8	36,812	38,404	– 1,592	43,084	46,711	– 3,627

TABLE 3.3—*Continued*

Revenue and Expenditure of States in First Plan

Part B States	1951–52 R	1951–52 E	1951–52 ±	1952–53 R	1952–53 E	1952–53 ±	1953–54 R	1953–54 E	1953–54 ±	1954–55 R	1954–55 E	1954–55 ±	1955–56 R	1955–56 E	1955–56 ±
1. Hyderabad	2,958	2,790	168	2,646	2,504	142	2,535	2,649	– 114	2,717	2,802	– 85	2,676	2,912	– 236
2. Madhya Bharat	1,107	1,089	18	1,136	1,181	– 45	1,405	1,241	164	1,450	1,371	79	1,698	1,847	– 149
3. Mysore	1,391	1,395	– 4	1,447	1,372	75	1,513	1,524	– 11	1,612	1,655	– 43	2,535	2,535	– 0·3
4. Pepsu	605	462	143	610	534	76	688	672	16	757	893	– 136	906	1,079	– 173
5. Rajasthan	1,522	1,547	– 25	1,790	1,569	221	1,851	1,798	53	2,193	2,004	189	2,397	2,513	84
6. Saurashtra	747	858	– 111	977	1,168	– 191	1.056	842	214	1,229	1,227	2	1,345	1,619	– 274
7. Travancore – Cochin	1,724	1,296	428	1,593	1,546	47	1,618	1,386	232	1,686	1,203	483	1,864	1,966	– 102
TOTAL	10,054	9,437	617	10,199	9,874	325	10,666	10,112	554	11,644	11,155	489	13,421	14,471	– 1,050
GRAND TOTAL	40,534	39,268	1,266	42,014	41,703	311	45,366	44,820	546	48,456	49,559	– 1,103	56,505	61,182	– 4,677

*Budget Estimates. †Difference due to rounding off.

[Source: Upto 1954–55, figures are from Reserve Bank of India Bulletins. For 1955–56, figures are from the Combined Finance and Revenue accounts.]

D

TABLE 3.4

Overall Position of States in First Plan

(Rs. Lakhs) (actuals)

Part A States	1951–52 O±	1951–52 O	1951–52 C	1952–53 O±	1952–53 O	1952–53 C	1953–54 O±	1953–54 O	1953–54 C	1954–55 O±	1954–55 O	1954–55 C	1955–56 O±	1955–56 O	1955–56 C
1. Andhra Pradesh	..	..	..	..	..	..	..	..	..	−705	800	86 (9)	−70	86	16
2. Assam	123	321	177 (267)	−276	177	292 (−391)	61	292	302 (51)	287	302	506 (83)	29	506	535
3. Bihar	−1,209	−241	−178 (−1,272)	221	−178	297 (−254)	271	296	651 (−84)	330	651	944 (37)	−805	944	139
4. Bombay	−608	−115	53 (−776)	134	29	308 (−145)	1,101	308	565 (844)	2,402	565	1,048 (1,919)	759	1,211	1,971
5. Madras	−1,742	−479	−362 (−1,859)	−212	−362	−66 (−508)	−416	−919	−1,335*	−87	714	631 (−4)	253	631	884
6. Madhya Pradesh	305	412	580 (+137)	−168	580	272 (140)	557	271	71 (757)	237	71	388 (−80)	86	388	474
7. Orissa	226	112	121 (−7)	110	121	237 (−6)	−50	237	190 (−3)	240	190	285 (145)	−392	285	−106
8. Punjab	−773	1,820	997 (50)	−403	997	544 (50)	−508	544	123 (−87)	684	123	908 (−101)	−207	908	701
9. Uttar Pradesh	−584	714	22 (108)	−752	22	756 (−1,486)	1,014	756	1,566 (204)	−729	1,566	491 (346)	906	491	1,397
10. West Bengal	−14	742	728	−169	728	759 (−200)	−395	759	364	542	364	906	803	906	1,708
TOTAL	−4,276	3,062	2,138 (−3,352)	−1,515	2,114	3,399 (−2,800)	1,635	2,544	2,597 (1,682)	3,201	5,346	6,193 (2,354)	1,362	6,356	7,719

TABLE 3.4—*Continued*

Overall Position of States in First Plan

Part B States	1951–52 O±	1951–52 O	1951–52 C	1952–53 O±	1952–53 O	1952–53 C	1953–54 O±	1953–54 O	1953–54 C	1954–55 O±	1954–55 O	1954–55 C	1955–56 O±	1955–56 O	1955–56 C
1. Hyderabad	198	1,203	1,401	-221	1,401	1,180	292	838	679 (-133)	- 25	679	577 (77)	102	577	679
2. Madhya Bharat	- 40	261	266 (-45)	81	266	308 (39)	175	308	341 (142)	300	341	394 (247)	14	394	408
3. Mysore	50	428	481 (-3)	-218	481	352 (-89)	565	352	559 (358)	243	559	625 (177)	-410	675	265
4. Pepsu	13	352	365	87	365	389 (87)	- 5	389	326 (58)	-216	326	214 (-104)	- 59	214	155
5. Rajasthan	-145	168	89 (-66)	-198	89	45 (-154)	- 31	45	4 (10)	30	4	36 (-2)	28	36	64
6. Saurashtra	14	160	41 (133)	-109	41	90 (-158)	- 59	90	101 (-70)	435	101	427 (109)	- 48	427	379
7. Travancore -Cochin	-593	514	312 (-391)	23	312	367 (-32)	- 75	367	218 (74)	268	218	510 (-24)	- 38	510	472
TOTAL	-503	3,086	2,955 (-372)	-555	2,955	2,731 (-331)	252	2,389	2,228 (413)	1,035	2,228	2,783 (480)	-411	2,833	2,422
GRAND TOTAL	- 4,779	6,148	5,093 (-3,724)	-2,070	5,069	6,130 (-3,131)	1,887	4,933	4,825 (2,095)	4,236	7,574	8,976 (2,834)	951	9,189	10,141

O± = Overall position. O = Opening Balance. C = Closing Balance. * Budget Estimate.
Minus figures in brackets indicate disinvestment and plus figures, investment, in funds and securities.
[Source: Reserve Bank of India Bulletin. For 1955–56 only, combined Finance and Revenue Accounts.]

(3) adverse seasonal conditions and other natural calamities; and (4) inauguration of the First Plan. Tables 3.3 and 3.4 will show the surplus and deficit in revenue account and in overall transactions during this period.

The effect of the states' mergers on the finances of provinces was generally unfavourable as revenue collections from the merged areas were less than the expenditure they entailed. According to the 1941 Census data the population of the nine Provinces (allowing for partition) was 228 million while the population of the Indian states merged with this population was 19·5 million. The addition to population was the highest for Bombay (85 lakhs), Orissa (51 lakhs) and Madhya Pradesh (28 lakhs). State governments were faced with the problem of gradually raising the administrative levels and improving social services in the merged areas. For 1950–51, the gap between revenue and expenditure in respect of territories merged with Bombay, Orissa and Madhya Pradesh was estimated at Rs. 540 lakhs, 100 lakhs and 45 lakhs respectively. The financial implications of state mergers were impressed upon the Union Government which, adopting the recommendations of the Indian States Finance Enquiry Committee, agreed to a restricted scheme of financial assistance equal to the excess of federal income over federal expenditure in the merged areas or half of the net income tax collection in the merged areas whichever was higher.

Although in the financial provisions of the Constitution both Part 'A' and Part 'B' states had the same rights and the same relations with the Union Government, in view of the disparities in the relative tax structures of these two classes of states, the rights of Part 'B' states were governed by the federal financial agreement entered into with them by the centre. For example, it was decided as a transitional measure to allow some of these states a period of time ranging up to five years in which to abolish internal customs.

The strain imposed by the First Plan on the finances of the states varied according to the size of the Plan in relation to the state of development of the resources of the states, the varying impact of different factors on the growth of tax revenues, unavoidable commitments of a non-developmental or emergency character, such as famine relief and rehabilitation of displaced persons and the measure of tax effort and borrowing undertaken by the respective state governments. While almost every state more or less attained the plan targets in financial terms, some of the states had to cope with special difficulties and were left with no balances at the end of the plan period. Bihar for example had to consistently face floods and drought, the direct expenditure on this account being Rs. 9·39 crores in 1954–55 and Rs. 7·42 crores in 1955–56. A sizeable part of the expenditure was met by the state government itself. By 1956–57 its development fund had to be written off and the overdraft in the consolidated fund was expected to rise from Rs. 16 crores at the end of 1955–56 to Rs. 29·5 crores at the end of 1956–57. Consequently the state was unable to make any contribution from current revenue for financing plan expenditure. West Bengal, on the other hand, though it had the problem of continued influx of refugess from East Pakistan, was not involved very much financially since its own share in the expenditure was relatively small. The budgets of Madras were affected by abnormal expenditure connected with cyclone damage. As at the end of March, 1956, the state had no invested balances, hardly any even in the sinking fund account. The state of Bombay however increased its cash balance and cash balance investment account between 1951–52 and 1956–57 and 'neither prohibition nor plan responsibilities left the state poorer'.

Thus the trends of the state finances during the First Plan varied from state to state. The Second Finance Commission which reviewed the trend of state finances between 1952–53 and 1955–56 was not particularly alarmed by the financial position of the states. The important conclusions of its review of state finances were: (1) The scale of devolution recommended by First Finance Commission was generally adequate for the normal expenditure of most states and allowed for many of them a sizable surplus after meeting the plan expenditure. (2) The tax effort among states during this period fell far short of the expectations of the Planning Commission. If they had raised the resources expected of them, some of the states which ran into deficit might not have done so. (3) The improvement under the principal heads of revenue was of the order of 5% per annum. It was due partly to the normal expansion of revenue, partly to increased receipts of land revenue in certain states owing to the abolition of Zamindari and partly to the additional taxes. Out of the additional revenue raised about 50% was accounted for by sales tax and taxation of motor spirits and tobacco and 20% by motor vehicles, passenger and carriage taxes. The balance was raised by a number of minor taxation measures. Taxation of land contributed very little and except in Uttar Pradesh, irrigation rates did not contribute any sizable amount. (4) The level of arrears of revenue and over-due loans became a matter of concern in some states. (5) The public debt of the states was increasing rapidly on account of the implementation of the Plan and a considerable part of it was likely turn out to be unproductive debt the cost of whose servicing would have to be met from general revenues. Efforts had to be made to ensure that irrigation, electricity, transport and commercial schemes yielded the maximum revenue so as to keep down the net burden of interest charges. (6) Expenditure was steadily rising as a result of development. Non-development expenditure rose at a pace somewhat in excess of that envisaged by the plan.

An important feature of the public finances of India during the First Plan was the investigation and report of the Taxation Enquiry Commission, 1953–54. The last enquiry into taxation had been conducted by the Indian Taxation Enquiry Committee nearly thirty years before. The Taxation Enquiry Commission of 1953–54 went into the federal, state and local finances in great detail though many of its recommendations were not implemented subsequently. It will not be possible for us to summarize the recommendations of the Commission here. But some of the trends noted by it are worth repeating. The Commission found that compared with other federations the field within which public revenues were raised and spent regionally was much wider in India, which testified to the greater importance of state governments in the federal system. It also found that there had been very little addition to the national tax effort over the last two or three decades in average terms. Revenues of local bodies in particular grew very slowly. The sharing by the state governments in the proceeds of central taxes and the institution of substantial grants by the central government for both general and specific purposes between them imparted a measure of elasticity to the state finances which was unknown to the provinces in the pre-war period. Even in the taxes which were both collected and retained by the states there was a greater degree of elasticity than was in evidence before the war. 'Central and State revenues' the Commission stated, 'really coalesce for purposes of the public finance of state governments and the old antagonism between central revenues and state revenues has largely disappeared. To this extent, an integrated treatment of Indian public finance becomes much more in order'.[1]

We now turn to the overall trends in state finances during the Second Five Year Plan. The reorganization of states in November 1956 necessitated readjustments of the finances of most of the states and also the allocation of assets (including cash balances) and liabilities. By this time the impact of the plan expenditure on the revenue

[1]P. 210, Report of the Taxation Enquiry Commission, 1953–54, Volume 1.

budget had already been felt and it was considered that certain relaxations of orthodox budgetary principles were necessary to keep pace with development. A circular was issued by the centre to the states suggesting certain re-classifications of expenditure between revenue and capital from 1956–57. We have already seen in this chapter what these changes were. The broad idea behind these changes was that at a time when strain on the revenue budgets of states was inevitable they should not be further strained by items which could be classified as capital. Thus works of a capital nature, the cost of which exceeded Rs. 20,000 as also smaller ones which form part of a scheme whose total cost exceeded Rs. 1 lakh were transferred to capital though until then they were being shown under capital expenditure.

It was suggested that grants to local bodies and other public institutions for works may also be debited to capital provided the expenditure was written back to revenue over a period of, say, fifteen years. The provision of repayment of both market loans and borrowing from the centre was to be shown wholly in the public debt section. When there was no real surplus available for repayment of debt and when borrowings were all for development purposes, it was argued, there was little justification for including such provision in the revenue budget. The net result of these changes was a fairly substantial relief to the revenue budget.[1] For example the relief to revenue budget for 1956–57 was estimated at Rs. 1·63 crores in the estimate of Bombay, Rs. 3·5 crores in the case of Madras, Rs. 2·25 crores for Travancore and Cochin, Rs. 1·92 crores in respect of Punjab and over Rs. 1 crore for Mysore.

In spite of the illusory relief to the revenue budget some states continued to run into heavy weather in capital waters. States like Bihar ended with a deficit in the overall transactions year after year. Plan grants began to lapse and one of the reasons was that states had no money to carry out the schemes and central assistance, under the recoupment system, had to be claimed after the expenditure. From 1958, therefore, a procedure was devised by which central assistance was released as a ways and means advance and was later adjusted with reference to performance.[2] The states did better in

[1]It has been argued that had the centre not suggested these changes, the true revenue position of the state governments would have been known and they would have obtained better treatment at the hands of the later Finance Commissions. See Madras Government's Memorandum to the Finance Commission, 1965.

[2]Described in Chapter IV.

additional taxation than they did during the First Plan and their plan performance in financial terms was also adequate.

In 1957 the Government of India announced a scheme of central assistance for the increase in emoluments of employees of state governments to be operative for four years from 1957–58 after which the Finance Commission would be expected to take care of the expenditure. It proposed a complicated formula (correspondence on entitlements is still going on in some states) according to which grants would be given for a portion of the expenditure on the increase of emoluments for 'low paid' employees. For employees whose emoluments after the increase would be Rs. 100/- but not Rs. 250/- the central government was prepared to assist the states by means of a loan. Following this, a number of states increased the emoluments of their employees. This was to be the first of a series of revision of emoluments which have considerably added to the commitments of the state governments. Invariably the central government used to raise the emoluments first by way of increase in the dearness allowance and the states had to follow suit sooner or later but with uncertainty regarding availability of central assistance. The revision of emoluments is still an uncertain factor in state budgeting.

There were some other changes of note during the Second Plan period. Since they will be discussed in the chapter on 'Union State Relations', we shall confine ourselves here to merely mentioning them:

(1) Classification of income tax paid by companies as corporation tax which meant the shrinkage of the divisible pool; which in turn had to be compensated till 1960–61 by *ad hoc* grants.

(2) The tax on railway passenger fares was merged with the basic rates and the states were given instead a grant-in-aid on the basis of the previous collections.

(3) Due to difficulties experienced in inter-state trade the Central Sales Tax Act was passed.

(4) In 1957 the states agreed to give up their sales taxes on certain commodities in lieu of additional excise duties to be levied by the centre. The states were compensated for the loss by the share-out of the proceeds of the additional excise on the basis of the recommendations of the Finance Commission.

Tables 3.5 and 3.6 depict the financial position of the states during the Second Plan in the Revenue Account as well as in overall transactions.

By the end of the Second Plan states' expenditure had gathered considerable momentum and though some states finished the Second Plan with a fairly comfortable cash balance the inherent instability of the state budgets was beginning to reveal itself. From 1961–62, the first year of the Third Plan, there began the phenomenon of the overdrafts of the states with the Reserve Bank of India. These became a permanent feature during the Third Plan and the overdrafts increased from year to year, as has been indicated in Chapter I itself. Apart from the pressure of the Plan expenditure, the added commitments by way of increased emoluments to employees, the Chinese aggression in 1962 and the hostilities with Pakistan in 1965 increased expenditure on civil defence and other matters especially in border areas. Important programmes like the programme of universal primary education also put additional strain on state finances. Throughout the Third Plan practically all the states had ways and means difficulties of a rather difficult nature. It was becoming increasingly clear that these ways and means difficulties reflected a fundamental disequilibrium in state finances.

Tables 3.7 and 3.8 depicts the financial position of the states during the Third Plan in the Revenue Account as well as in overall transactions.

By the time of the formulation of the Fourth Plan the finances of the state governments have become matters of concern. The centre has issued certain broad guide lines to state governments concerning their overdrafts. These are:

(1) Budgets must be balanced as far as possible.

(2) Expenditure must be directly linked to a realistic appraisal of resources in sight.

(3) Expenditure not provided in the budget should be avoided.

(4) There should be periodical review of expenditure and the impact of economy measures so as to prevent the situation getting out of control necessitating overdrawals from the Reserve Bank of India.

These guide lines were just a part of a series of exhortations on the part of the centre to the states in recent years. The latest scheme of the centre, probably unlikely to be implemented, is that the limits for authorized overdrafts for the State Governments will be raised, but if those limits are exceeded, the State Governments in question will have to repay the excess within three weeks—otherwise the

TABLE 3.5

Revenue and Expenditure of States in Second Plan

(Rs. Lakhs) (actuals)

States	1957–58			1958–59			1959–60			1960–61		
	R	E	±	R	E	±	R	E	±	R	E	±
1. Andhra Pradesh	6,128	5,406	722	6,619	6,209	410	8,043	7,312	731	8,360	8,329	31
2. Assam	2,856	2,756	100	3,105	2,785	320	3,403	3,184	219	3,269	3,719	—450
3. Bihar	4,871	5,867	-996	6,005	6,071	- 66	6,747	6,798	- 51	7,861	7,113	748
4. Bombay	12,652	11,445	1,207	7,805	6,940	865	..	..	..	..	..	..
5. Gujarat	..	..	..	..	..	..	..	..	..	5,201	5,068	133
6. Jammu & Kashmir	857	704	153	1,078	862	216	1,254	1,076	178	1,502	1,147	355
7. Kerala	2,747	2,888	- 141	3,432	3,473	- 41	3,712	3,986	- 274	4,420	4,496	- 76
8. Madhya Pradesh	5,025	4,843	182	5,830	5,259	571	6,007	5,604	403	6,825	6,296	529
9. Madras	6,200	5,859	341	6,934	6,774	160	8,041	7,988	53	9,126	9,032	94

TABLE 3.5—*Continued*

Revenue and Expenditure of States in Second Plan

(Rs. Lakhs)

States	1957–58			1958–59			1959–60			1960–61		
	R	E	±	R	E	±	R	E	±	R	E	±
10. Maharashtra	..	..	..	..	..	..	14,748	14,675	73	11,543	11,724	-181
11. Mysore	4,333	3,829	504	5,207	4,216	991	5,596	4,919	677	6,206	5,979	227
12. Orissa	2,148	2,292	-144	2,669	2,508	161	2,769	2,785	- 16	3,448	3,456	- 8
13. Punjab	4,217	3,464	753	4,841	4,036	805	5,591	4,483	1,108	6,115	5,626	489
14. Rajasthan	3,011	3,074	- 63	3,322	3,519	-197	3,859	3,956	- 97	4,315	4,465	-150
15. Uttar Pradesh	9,644	9,092	552	10,760	10,651	109	11,873	11,734	139	13,514	13,110	404
16. West Bengal	6,730	6,920	-190	7,944	7,841	103	9,056	8,480	576	9,476	9,177	299
TOTAL	71,419	68,439	2,980	75,551	71,144	4,407	90,699	86,980	3,719	101,181	98,737	2,444

[Source: Reserve Bank of India Bulletins.]

TABLE 3.6

Overall Position of States in Second Plan

(Rs. Lakhs)

States	1957–58			1958–59			1959–60			1960–61		
	O±	OB	CB	O±	OB	CB	O±	OB	CB	O±	OB	CB
1. Andhra Pradesh	97	184	300 (−19)	222	300	595 (−73)	−118	596	152 (326)	−10	152	150 (−8)
2. Assam	−149	222	73	−87	73	−14	95	−14	81	−849	81	−768
3. Bihar	−406	−261	−667	−91	−667	−758	169	−758	−589	240	−589	−349
4. Bombay	712	714	1,169 (257)	1,447	1,079	505 (2,021)	..	..	..	..	..	..
5. Gujarat	..	..	..	..	..	..	..	..	..	−475	474	−231 (230)
6. Jammu & Kashmir	−27	208	181	16	180	196	−42	196	154	26	154	180
7. Kerala	−72	99	95 (−68)	−4	95	174 (−83)	−397	174	−150	2	−150	−78 (−70)
8. Madhya Pradesh	−423	55	−224 (−144)	−173	−224	−377 (−20)	395	−377	145 (−127)	−96	145	141 (−92)
9. Madras	−279	243	−36	388	−36	328 (24)	728	328	442 (614)	−41	442	552 (−151)
10. Maharashtra	..	..	..	..	..	..	−2,473	501	−1,551 (−421)	−1,918	−1,548	−387 (−3,345)

TABLE 3.6—*Continued*

Overall Position of States in Second Plan

(Rs. Lakhs)

States	1957–58			1958–59			1959–60			1960–61		
	O±	OB	CB	O±	OB	CB	O±	OB	CB	O±	OB	CB
11. Mysore	295	754	1,012 (37)	1,003	1,012	287 (1,728)	276	287	143 (420)	-1,199	143	142 (-1,198)
12. Orissa	- 29	-402	-448 (17)	792	-448	185 (159)	648	185	-139 (972)	-131	-139	16 (-286)
13. Punjab	- 37	496	314 (145)	271	314	837 (-252)	603	837	594 (846)	-1,870	594	121 (-1,397)
14. Rajasthan	- 57	21	37 (-73)	17	37	54	103	54	81 (76)	- 32	81	81 (-32)
15. Uttar Pradesh	-468	-141	-559 (-50)	261	-559	-298	234	-298	- 64 (71)	1,136	- 64	123 (949)
16. West Bengal	-901	571	297 (-627)	1,309	297	88 (1,518)	848	88	229 (707)	379	229	220 (388)
TOTAL	-1,744	2,763	1,544 (-525)	5,371	1,453	1,802 (5,022)	1,069	1,298	-472 (2,870)	-4,838	1	- 87 (-5,012)

[Source: Reserve Bank of India Bulletins.]

Note: For explanation, see Table 3.4.

TABLE 3.7

Revenue and Expenditure of States in Third Plan

(Rs. Lakhs) (actuals)

States	1961–62			1962–63			1963–64			1964–65			1965–66‡		
	R	E	±	R	E	±	R	E	±	R	E	±	R	E	±
1. Andhra Pradesh	8,460	8,903	-487	10,910	10,456	454	13,172	12,107	1,065	13,780	13,336	444	14,772	16,123	-1,351
2. Assam	3,943	4,152	-209	4,268	4,249	19	4,686	5,165	-479	5,567	6,127	-560	7,063	7,739	-676
3. Bihar	7,848	7,971	-123	9,098	8,110	988	9,724	8,456	1,268	10,823	8,692	2,131	12,109	11,992	117
4. Gujarat	6,200	6,367	-167	8,215	6,976	1,239	8,392*	8,503	-111	9,659*	9,292	367	10,498*	10,698	-200
5. Jammu & Kashmir	1,863	1,560	303	2,038	2,015	23	2,352	2,289	63	2,310	2,534	-224	2,740	3,322	-582
6. Kerala	4,696	5,337	-641	5,926	5,958	- 32	6,436	6,302	134	7,480	6,817	601	8,070	8,177	-107
7. Madhya Pradesh	7,448	7,667	-219	7,719	8,124	-405	10,060	9,332	-285	10,569	10,124	445	11,797	12,107	-310
8. Madras	9,134	10,095	-961	11,189	11,573	-384	12,713	12,736	157	14,513	14,255	258	15,541	16,328	-787

TABLE 3.7—*Continued*

Revenue and Expenditure of States in Third Plan

States	1961–62			1962–63			1963–64			1964–65			1965–66		
	R	E	±	R	E	±	R	E	±	R	E	±	R	E	±
9. Maharashtra	11,677	12,429	- 752	14,593	14,272	322	18,361	16,494	1,867	19,843	19,933	- 90	21,556	23,757	- 2,201
10. Mysore	6,594	7,061	- 467	7,670	7,878	- 208	8,853	8,523	330	9,557	9,266	291	10,508	11,272	- 764
11. Orissa	4,278	5,799	- 1,521	5,894	6,266	- 372	6,617	6,889	- 272	7,183	7,606	- 423	8,169	8,846	- 677
12. Punjab	7,018	5,954	1,064	8,450	7,504	946	10,550	9,273	1,277	11,447	10,141	1,306	12,199	11,457	742
13. Rajasthan	4,528	5,107	- 579	5,693	5,516	177	6,587	6,664	- 77	7,273	7,710	- 437	9,657	10,066	- 409
14. Uttar Pradesh	13,660	13,602	58	16,151	16,017	134	18,141	17,007	1,134	19,653	19,326	327	21,923†	22,512	- 589
15. West Bengal	10,045	10,123	- 78	10,572	11,174	- 602	12,376	11,506	870	13,903	13,314	589	16,493	16,923	- 430
TOTAL	107,349	112,127	- 4,779	128,386	126,088	2,299	149,020	141,246	7,774	163,498	158,473	5,025	183,095	191,319	- 8,224

*Excludes receipts from Revenue Reserve Funds of Rs. 585 lakhs, Rs. 561 lakhs and Rs. 526 lakhs for 1963–64, 1964–65, 1965–66 respectively.
†Excludes receipts from Revenue Reserve Fund, Rs. 589 lakhs.
‡Revised estimates.

[Source: Reserve Bank of India Bulletins.]

TABLE 3.8

Overall Position of States in Third Plan

(Rs. Lakhs)

States	1961–62			1962–63			1963–64			1964–65			1965–66*		
	O±	OB	CB	O±	OB	CB	O±	OB	CB	O±	OB	CB	O±	OB	CB
1. Andhra Pradesh	– 14	150	138 (–2)	369	138	246 (261)	645	246	797 (91)	..	797	613 (184)	– 2,915	613	– 2,302
2. Assam	112	–768	–656	510	–656	–146	54	–146	– 92	556	– 92	464	– 1,032	–678	– 1,710
3. Bihar	–661	–349	– 1,010	801	– 1,010	–209	743	–209	545 (–11)	47	545	295 (297)	–843	295	–548
4. Gujarat	330	–230	207 (–107)	102	203	810 (595)	– 1,255	810	–409 (–36)	42	–408	–626 (260)	–556	–626	94 (–1,276)
5. Jammu & Kashmir	62	180	242	– 8	242	234	29	234	264 (–1)	92	264	362 (–6)	–260	362	102
6. Kerala	–392	– 78	–404 (–66)	350	–413	– 85 (22)	–543	– 85	–642 (14)	1,054	–642	416 (–4)	–440	416	– 26 (2)
7. Madhya Pradesh	–438	141	–281 (–16)	–428	–281	–752 (43)	–285	–751	–977 (–59)	592	–977	–339 (–46)	– 2,992	–339	– 3,366 (35)
8. Madras	–741	552	244 (–433)	–244	252	25 (–17)	157	25	183 (–1)	551	183	734	–707	735	40 (–12)

TABLE 3.8—*Continued*

Overall Position of States in Third Plan

(Rs. Lakhs)

States	1961–62			1962–63			1963–64			1964–65			1965–66		
	O±	OB	CB	O±	OB	CB	O±	OB	CB	O±	OB	CB	O±	OB	CB
9. Maharashtra	870	-387	510 (-27)	1,884	514	88 (2,310)	2,435	88	-939 (3,462)	492	-940	-214 -234)	-6,051	-214	60 (-6,325)
10. Mysore	-601	142	834 (-1,293)	-821	834	-17 (30)	264	-17	432 (-185)	-26	432	622 (-216)	-2,491	622	-1,088 (-781)
11. Orissa	-427	'16	-216 (-195)	209	-216	-130 (123)	-1,394	-130	-828 (-696)	-14	-828	-841 (-1)	888	-842	46
12. Punjab	-718	121	-384 (-213)	568	-384	424	-329	424	95	-863	95	-510 (-258)	-1	-510	-581 (70)
13. Rajasthan	-907	81	56 (-882)	-124	56	71 (-139)	369	783	1,225 (-73)	16	432	76 (372)	-340	76	109 (-373)
14. Uttar Pradesh	-233	123	373 (-483)	-277	373	96	-12	96	84	355	84	439	-965	438	62 (-589)
15. West Bengal	10	220	170 (160)	-1,244	170	305 (-1,379)	-1,426	+304	101 (-1,223)	983	101	189 (895)	-187	189	22 (-20)
TOTAL	-3,748	-86	-177 (-3,657)	2,747	-178	960 (1,609)	-548	1,672	-161 (-1,285)	3,877	-954	1,680 (1,243)	-18,892	-537	-9,086 (-9,269)

*Revised Estimates.

[Source: Reserve Bank of India Bulletins.]

Note: For explanation, see Table 3.4.

E

Reserve Bank will have the right not to honour their cheques until the excess is repaid.[1]

The salient features of the trends in state finances during the first three Plans may now be summarized. In the rest of the chapters we shall be referring repeatedly to these features.

(1) There was a phenomenal growth of the size of operation of the state governments in the revenue as well as capital account. This was particularly due to the implementation of the Plans and increased development expenditure, though non-development expenditure has also exhibited a growth of its own. There was a shift towards development expenditure particularly on the capital account. The increase in the size of operation of the states was made possible largely by central assistance and partly by the deepening and widening of the tax structure of the states. There was considerable buoyancy of the state revenues along with development and the states also put forward substantial efforts of additional taxation, except during the First Plan. But the financial position of the states collectively and individually ran into ways and means difficulties not of a passing nature but indicative of a fundamental disequilibrium.

(2) Central control and assistance have come to occupy a very prominent place in the finances of state governments. A dualism in central assistance has developed and there has been a certain overlapping of the functions of the Finance Commission and the Planning Commission. The central patterns of assistance for plan schemes have been targets of criticism by the states.

(3) The volume of debt of the state governments has increased enormously. The interest charges have mounted very considerably and the problem of servicing the debt which has been spent in productive as well as non-productive ways, has become a matter of great concern.

We shall stop our review of the overall trends of state finances at this stage and go into the various aspects of public finances in some detail and build up a perspective after which, in the concluding chapter, we can again take up the thread of the overall position of state finances.

[1]See *Capital*, 23rd March, 1967, page 558. *Capital* commented: 'In this scheme lie the seeds of a first class political row, especially now that non-Congress Governments have come to power in several states. And, to make matters worse, the scheme itself is fundamentally unfair, wrong-headed and incomplete—because it imposes no corresponding obligation of financial discipline on the centre'.

CHAPTER IV

FINANCING THE STATE PLANS

Planning in India is a subject of various ramifications of which we are concerned here only with those aspects which impinge directly on state finances. A discussion of these aspects is quite essential to the scheme of our study since the implementation of the plans has had a vital impact on state finances. We shall concern ourselves only with the size of the state plans, their composition in terms of schemes and in terms of various types of resources and the impact that planning has had on the finances of state governments. Before we proceed to consider these aspects we may do well to bear in mind certain classifications of expenditure which are relevant to our appreciation of the finances of state governments.

We have seen that the budget of a state government consists, inter alia, of a revenue account as well as a capital account comprised wherein are capital expenditure outside revenue account and loans. The concept of plan outlay is however different in the sense that it does not fit in exactly with the divisions of the budget. A plan ultimately boils down to a list of schemes on which expenditure has to be incurred from various sections of the budget. Thus plan outlay consists of revenue expenditure as well as capital expenditure outside the revenue account as well as loans to various parties. A particular plan scheme may involve expenditure both in the revenue account and in the capital account. For example a scheme like appointment of a teacher for primary education is a scheme involvping payment of salaries wholly debitable to the revenue account. A scheme like a subsidy for purchase of pesticides involves expenditure wholly from the revenue account. On the other hand a scheme like the establishment of state seed farms and stores involves staff expenditure on revenue account and capital expenditure on capital account. A scheme of soil conservation will involve revenue expenditure for the staff employed as well as expenditure by way of loans in the capital account for the quantity of work done on behalf of land owners. Thus the expenditure proposed and incurred on a plan scheme cannot be gleaned straightaway from the budget since the expenditure will be shown in different sections of the budget depending on the nature of the scheme. This is one of the accounting and presentational problems facing the state governments which they have to solve. Madras, for instance, publishes a separate book showing the schemes under its plan and their outlays under 'revenue', 'capital' and 'loan' and also their correlation in the detailed budget estimates.

It will also be realized that plan outlay is only part of the total budget of a state government. This budget will comprise of what are known as non-development items of expenditure like expenditure on law and order, general administration, jails, expenditure on collection staff, debt services and so on. It will also comprise of what may be called developmental expenditure consisting of expenditure on such services as education, medical, public health and so on. The plan outlay for any particular year is even narrower in scope than 'development expenditure'. Ultimately speaking, the plan outlay is nothing but the outlay under various heads of the budget on schemes listed as part of the plan of the state. Expenditure on schemes not so listed even though it is of a developmental nature would be left out. If a state takes on a developmental programme outside the plan it will be developmental expenditure, no doubt, but non-plan expenditure. Again a convention has developed whereby all staff and maintenance expenditure of schemes taken up under a particular plan is not reckoned as plan expenditure in the subsequent plan but is treated as 'committed' expenditure which the state governments are committed to meet themselves without any plan assistance.

It will thus be seen that there are various possible classifications of government expenditure: first, on the usual budget lines as revenue, capital and loan, secondly as development and non-development expenditure and thirdly as plan and non-plan expenditure. Diagram 4.0 attempts to illustrate the differences we have explained above.

We may now turn to the substantive aspects of plan financing. 'Economic and social planning', we have already seen, is a concurrent subject. The nation as a whole is vitally interested in economic development and the centre has a great stake in it but it is the states that have to executive a large number of schemes necessary for planned development. In the context of inadequate state finances and the need for large outlays on development schemes, central aid assumes a crucial role in the implementation of the Plans. Planning ultimately becomes, at the stage when the desired rates of growth are transmuted into desired outlays in various sectors, just a list of schemes in the central sector, the state sector and the

4.0 DIAGRAM SHOWING CLASSIFICATIONS OF EXPENDITURE

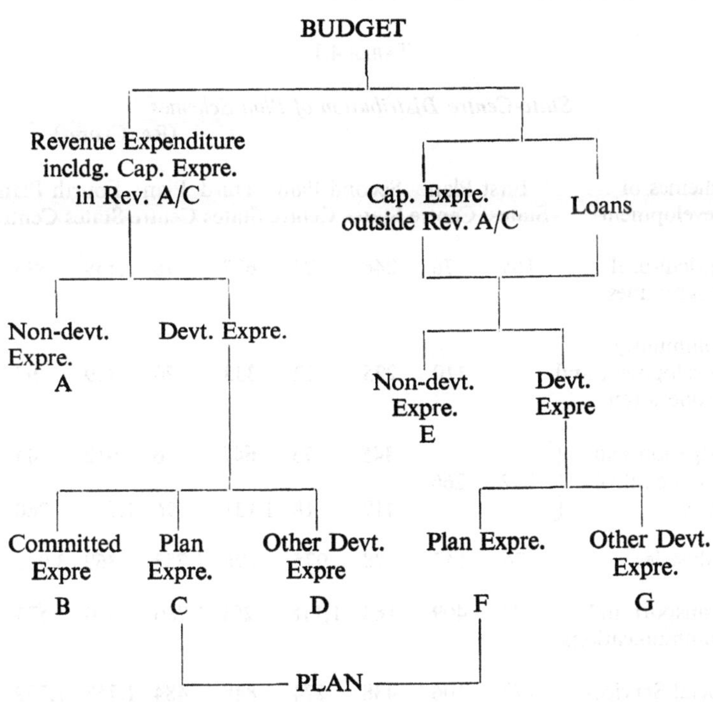

C + F = Plan expenditure

B + C + D + F + G = Development expenditure

A + ...G = Total expenditure

Note: Non-development expenditure includes collection of taxes, debt services, civil administration, miscellaneous, famine and others.

Development expenditure includes education, medical and public health, agriculture, veterinary and cooperation,, irrigation, electricity schemes, civil works, industries and supplies and others.

private sector. Speaking in constitutional terms the plan in the central and state sectors becomes nothing but a list of schemes on the subjects in the state list, the concurrent list and the central list. The following table will show the relative emphasis in planning on schemes in the central and state sectors.

TABLE 4.1

State-Centre Distribution of Plan Schemes

(Rs. Crores)

Schemes of Development	First Plan States	First Plan Centre	Second Plan States	Second Plan Centre	Third Plan States	Third Plan Centre	Fourth Plan States	Fourth Plan Centre
Agricultural Programmes	165	76	246	24	637	78	1,559	335
Community Development and Cooperation	—	110	225	22	338	20	359	97
Irrigation and Flood control	292 (Irrigation and Flood control and Power)	266 (Irrigation and Flood control and Power)	345	75	647	6	912	44
Power			419	18	1,131	96	1,717	260
Industries	26	147	98	975	196	1,753	369	3,922
Transport and Communications	79	409	163	1,118	291	1,786	380	2,575
Social Services	223	106	438	274	848	484	1,753	1,259
Other Programmes	11	127	48	51	67	41	24	44
TOTAL	796	1,241	1,982	2,557	4,155	4,264	7,073	8,536

[Source: Five Year Plans: The figures for the First and Fourth Plans are estimates; for the Second and Third Plans anticipated final figures as reported in the succeeding Plans.

It will be seen that the centre's investments have been largely towards industrial development and development of transport and communications. The state outlays are directed towards irrigation, power, agriculture and various social services.

It is needless to emphasize that planning is a field where close co-operation and co-ordination is necessary between the centre and the states. A plan itself is the product of the combined needs and the combined resources of the centre and the states. The process of plan formulation begins with the working groups at the central level for various broad items of development, (which have state representatives) which begin to list out schemes and outlays within the limits of the desired rates of growth as dictated by perspective planning. After the central working groups begin to function, similar working groups begin their deliberations in the states. As against these functional groups there are also territorial groups namely an inter-departmental group in each state[1] and the Planning Commission at the Centre. The pace is of course set by the Planning Commission and the functional working groups of the centre with which the state working groups, both territorial and functional, fall in line. Once the Plan is formulated and is put into action, annual plans are drawn up (this is from the middle of the Second Plan) indicating the outlays on various schemes for any particular financial year. Here again states have discussions with the functional working groups and the Planning Commission when the performance of the previous year is also reviewed. There is thus a considerable degree of co-operation and co-ordination between the centre and the states.[2] We shall be adverting in a later chapter to these consultations as forming part of centre-state relations.

The next important question is of course what decides the size of a state plan. We can approach this problem from the point of view of needs and the availability of physical resources as well as from the point of view of financial resources, the ultimate size of the plan being a resultant of these considerations. We shall consider later in the chapter the financial structure of the state plans. We may first look at the composition of state plans in terms of the relative emphasis on schemes laid by various states. Though overall magnitudes by themselves convey little meaning, unless they are shown in relation to some other magnitudes, we may begin by presenting the size of the state plans of various state governments during the first three plans.[3]

[1]Some states have also Planning Boards, consisting of Ministers, officials and non-officials.

[2]For a detailed description of the formulation and implementation of plans, see A. H. Hanson, *The Process of Planning*, chapter X. (Oxford University Press, 1966). This book is essential reading for all interested in Indian Planning.

[3]Fourth Plan figures are not yet available.

TABLE 4.2

Plan Outlays of States

(Rs. Crores)

	First Plan	Second Plan	Third Plan	
			Estimate	Actual
1. Andhra Pradesh	108	175	305	351
2. Assam	28	51	120	133
3. Bihar	102	166	337	317
4. Gujarat	224*	143	235	237
5. Jammu and Kashmir	13	25	75	63
6. Kerala	44	76	170	180
7. Madhya Pradesh	94	145	300	283
8. Madras	85	167	291	342
9. Maharashtra	*	207	390	425
10. Mysore	94	122	250	246
11. Orissa	85	85	160	219
12. Punjab	163	148	231	244
13, Rajasthan	67	99	236	204
14. Uttar Pradesh	166	227	497	530
15. West Bengal	154	145	250	299
TOTAL	1,427	1,981	3,847	4,073

[Source: p. 89, Third Five Year Plan and Reserve Bank of India Bulletin, May, 1966.]

*Figure for Bombay shown against Gujarat.

We cannot fully appreciate the figures we have given unless we study them in relative terms but one thing strikes the eye, namely, the various levels of outlays of different states. The levels of performance also vary though, by and large state governments have been able to reach their financial targets in the Plans. But how exactly the size of a state plan is arrived at is a matter on which various considerations will no doubt play their part and no easy formula is possible.[1] According to the Third Plan 'in determining the plan of each state consideration was given to its needs, problems, past progress and lags in development, likely contribution to the achievement of the major national targets, potential for growth and the contribution in resources which the state could make towards its development programmes. In assessing needs and problems such factors as population, area, levels of income and expenditure, availability of

[1]See also Hanson, op. cit.

certain services, for example, roads, schools, hospitals, extent of commitments carried over from the Second Plan, commitments on account of large projects or special programmes and the state of technical and administrative services are available were taken into account. Care was also taken to see that states whose resources were unavoidably small did not have to limit development to a scale which was altogether insufficient, merely because of paucity of resources. At the same time states which were able to make a larger effort in mobilizing their own resources could undertake development on an appropriate scale'.[1]

It will be interesting to note which groups of schemes have received emphasis in various states. Such a comparison over the four plan periods would be useful but due to states' reorganization we have to leave out the First Plan and due to non-availability of figures at the moment of writing we have to leave out the Fourth Plan. Table 4.3 will show the outlays under different heads of development by various states during the Second and Third Plans.

Before we begin interpreting this table we should remember that what it indicates is only the quantitative emphasis in terms of financial allocation and not the qualitative emphasis. Schemes such as irrigation and power absorb large outlays whereas certain schemes, say, in agriculture or family planning, may not require substantial outlay even though the states may attach equal or greater importance to them. Even so, the financial allocation among different schemes by different states offers an interesting study of the relative emphasis in different states on different programmes. Forgetting for a moment the absolute size of the plan of a state in relation to its needs and in relation to the plan of the other states, the inter se allocation reveals a fair amount of identity of approach (no doubt brought through the centre) to various development schemes. Expenditure on schemes of agriculture, community development and co-operation has been of the order of 20% to 25% in many states, the highest being 30%, in Uttar Pradesh and the lowest being 19% in Punjab and Jammu and Kashmir. The expenditure on irrigation (other than tanks and wells) has been of the order of 20%, the highest being 32% in Rajasthan and the lowest being 8% in Assam and 9% in Madras. It would be rash to conclude particularly in respect of schemes of irrigation and power that a lower outlay means lesser emphasis. The outlay under major irrigation depends on the natural resources and their economic availability as well as the stage of development already reached. In Rajasthan and Andhra

[1]P. 60, Third Five Year Plan, Planning Commission.

TABLE 4.3

Scheme Outlays in the Second and Third Plans
(Percentages of Total Outlay)

States	Agriculture, Community Development & Cooperation	Irrigation	Power	Industries	Transport and Communications	Social Services	Others (Residual)
(1)	(2)	(3)	(4)	(5)	(6)	(7)	(8)
1. Andhra Pradesh	23	28	20	6	3	18	2
2. Assam	21	8	18	8	10	34	1
3. Bihar	27	19	20	4	7	23	neg.
4. Gujarat	25	23	16	3	12	20	1
5. Kerala	23	10	27	9	6	24	1
6. Maharashtra	27	16	20	3	9	24	1
7. Madhya Pradesh	26	16	23	3	6	22	4
8. Madras	20	9	37	8	4	22	neg.
9. Mysore	24	18	25	6	7	20	neg.
10. Orissa	23	21	23	4	6	20	3
11. Punjab	19	20	30	5	6	18	2
12. Rajasthan	20	32	16	4	7	20	1
13. Uttar Pradesh	30	11	23	5	6	24	1
14. West Bengal	21	11	12	13	9	32	2
15. Jammu and Kashmir	19	16	14	11	15	20	5
All States	24	19	22	6	7	22	neg.

[Source: Third Five Year Plan Pp. 740–743. The Second Plan figures are anticipated final outlays and the Third Plan figures are estimates as at the beginning of the Plan.]

Note: (1) 'Agriculture, community development and cooperation' includes schemes under agricultural production, minor irrigation, soil conservation, animal husbandry, dairying and milk supplies, forests, fisheries, warehousing, marketing and storage, cooperation, community development and Panchayats.

(2) 'Irrigation' includes flood control but not tank or well irrigation. It includes only 'major' and 'medium' irrigation projects.

(3) 'Industries' includes large and medium industries, mineral development and village and small industries.

(4) 'Transport and communications' includes roads, road transport and tourism.

(5) 'Social services' includes general education and cultural programmes, technical education, health, housing, welfare of backward classes, social welfare, labour and labour welfare.

Pradesh the figures are high no doubt due to major projects like Bhakra Nangal[1] and Tungabadra. The terrain of Assam probably precluded major irrigation projects. In Madras the comparatively low outlay is due to the fact that the scope for major projects has been exhausted and the state itself depends to a large extent on minor irrigation.

Coming to power generation, states have allocated about 22% of their plan outlay for power. States like Madras and Punjab have allocated as much as 37% and 30% respectively whereas West Bengal has allocated only 12% and Gujarat and Rajasthan only 6%. Here again the outlay will be depending on the potentialities for power generation, the progress already achieved and the extent to which power generation is done through private licensees (which perhaps explains the low outlay on power, in states like West Bengal, Gujarat and Maharashtra).

The outlay on industries has been of the order of 6% only on an average. The highest is in Bengal which has 13%. Transport and communications, which for the states would mainly mean road development, absorbed only about 7% of the allocation; the highest being 12% in Gujarat and 10% in Assam and the lowest being 3% in Andhra Pradesh. About 22% of the outlay is absorbed by the social services and there is a fair measure of uniformity in this outlay. But the outlay in Assam and West Bengal, 34% and 32% respectively is noticeably high whereas the outlay in Andhra Pradesh and Punjab is only 18%.

We thus see that there is a fair measure of uniformity in allocation of their outlays by the states among various schemes though there are also variations, depending on local conditions and other factors. We should also remind ourselves before we turn away from this aspect, that the interpretation of the table can have only limited validity and cannot necessarily lead to a judgment on the question of the balanced development of various states.

We may now turn to what we may call the financial composition of a state plan. As an illustration, we may give in Table 4.4 the estimates of the Planning Commission of the financial resources of the state plans for the Third Plan period. It will be seen that the estimation of financial resources runs in terms of both revenue and capital accounts. The resources of the states on the revenue account are confined to the balance from current revenues (after meeting non-plan revenue expenditure) and

[1]A part of Rajasthan is benefited by the Bhakra Nangal Project which is in Punjab state.

TABLE 4.4

States' Resources for the Third Plan

(Rs. Crores)

	Andhra Pradesh	Assam	Bihar	Gujarat	Jammu & Kashmir	Kerala
1. Balance from current revenues at 1960–61 rates of taxation	14·8	7·0	19·8	10·2	8·0	10·6
2. Loans from the public (net)	40·0	9·0	23·0	30·0	..	23·0
3. Share of small savings	17·5	12·0	42·5	39·2	2·3	8·0
4. Unfunded debt (net)	3·2	2·8	8·0	4·0	1·7	3·0
5. Balance of miscellaneous capital receipts over non-plan disbursements	—29·7	—17·2	—36·6	—6·4	—8·2	—19·2
6. Contribution of enterprises	6·2	3·4	12·3	11·0	1·2	9·6
7. Withdrawal from cash and other reserves	..	..	..	5·0	..	..
8. Resources without taking into account additional taxation	52·0	17·0	69·0	93·0	5·0	35·0
9. Additional taxation	53·0	16·0	50·0	29·0	8·0	23·0
10. Total resources after taking into account additional taxation	105·0	33·0	119·0	122·0	13·0	58·0
11. Size of plan	305·0	120·0	337·0	235·0	75·0	170·0

	Madhya Pradesh	Madras	Maharashtra	Mysore	Orissa	Punjab	Rajasthan	Uttar Pradesh	West Bengal	Total
1.	—7·7	—30·2	30·6	9·8	—21·1	14·0	0·5	—43·5	11·3	34·1
2.	25·0	52·9	36·9	14·7	21·0	16·0	20·0	35·0	27·6	374·17
3.	17·0	20·0	66·8	10·0	8·5	35·0	10·0	50·0	38·5	377·3
4.	2·3	5·9	9·0	5·5	1·5	3·9	5·5	17·3	5·5	79·1
5.	—1·4	—33·7	15·3	..	—6·7	—28·3	10·00	—30·6	—40·6	—233·3
6.	9·8	35·1	9·4	13·0	1·8	16·4	2·0	9·8	7·7	148·7
7.	..	6·0	..	15·0	..	..	..	..	..	26·0
8.	45·0	56·0	168·0	68·0	5·0	57·0	48·0	38·0	50·0	806·0
9.	48·0	45·0	52·0	42·0	23·0	40·0	32·0	109·0	40·0	610·0
10.	93·0	101·0	220·0	110·0	28·0	97·0	80·0	147·0	90·0*	1,416·0
11.	300·0	291·0	390·0	250·0	160·0	231·0	236·0	497·0	250·0	3,847·0

*Revision of this amount to Rs. 133 crores was later suggested by the State Government.

[Source: Third Five Year Plan, pp. 118 and 89.]

on the capital account to the share of small savings, net unfunded debt and miscellaneous capital receipts and withdrawal of cash and other reserves. There are two essential decisions involved in arriving at the size of a state plan, namely, the quantum of additional taxation and the quantum of central assistance by way of grants and loans. Central assistance will proceed basically on the patterns of assistance for various schemes taken up in the state plans. Additional taxation has to be ultimately fixed with reference to what is considered feasible in a particular state. To the extent that central assistance is based on patterns it can be determined only after the size of a plan and its composition are themselves determined. To this extent the determination of a size of a state plan has an element of vagueness and arbitrariness about it. It is not clear on what basis the relative levels of additional taxation or central aid are fixed. Apparently the ultimate decision is rather on an ad hoc basis. It is not also clear whether the revenue implication of state plans are fully analysed at the stage of the formulation of plans. In the absence of any published material on this point one has only to take it that a great deal more of attention to the balancing of revenue budget will be necessary while formulating the plans. The additional taxation expected of the states varies quite considerably from state to state and this would immediately raise the question whether central aid is fair to all the states. The tables 4.5 and 4.6 will show a comparison of level of taxation in relation to the revenue component of the plan as well as in relation to the size of the state revenues at the beginning of the plan.[1]

Additional taxation as well as central assistance have therefore important parts to play. We shall be considering both these aspects in a later chapter in detail.

A mention should be made of the mechanism of central aid. Broadly speaking, there are two main types of schemes for which central aid is available namely the centrally assisted schemes and centrally sponsored schemes. These latter schemes are fewer in number than centrally assisted schemes and the outlays are also less. The grant/loan element is usually much higher in these schemes. The original idea must have been to bring under this head schemes in which the central departments were particularly interested but the state governments may or may not be equally interested either

[1]What we are adopting in presenting these tables is some kind of an *ad hoc* approach. For definitive conclusions a rigorous analysis is needed. For an attempt to assess the 'maladjustment' of tax efforts of States, see K. V. S. Sastri, *Federal-State Fiscal Relations in India*, Chapter II (Oxford University Press, 1966).

TABLE 4.5

Taxation and Revenue Component in Third Plan

(Rs. Crores)

State	Additional Taxation Target	Plan grant	Total revenue component	Plan
1. Andhra Pradesh	53	20	73	305
2. Assam	16	25	41	120
3. Bihar	50	53	103	337
4. Gujarat	29	35	64	235
5. Jammu and Kashmir	8	12	20	75
6. Kerala	23	20	43	170
7. Madhya Pradesh	48	33	81	300
8. Madras	45	33	78	291
9. Maharashtra	52	45*	97	390
10. Mysore	42	10	52	250
11. Orissa	23	30	53	160
12. Punjab	40	18	58	231
13. Rajasthan	32	28	60	236
14. Uttar Pradesh	109	53*	162	497
15. West Bengal	40	57	97	250
TOTAL	610	472	1,082	3,847

[Source: Third Plan and Report of the Finance Commission, 1961. It is better to explain in some detail the manner in which this table has been worked out, as no figures of the revenue component of the plan are published. But this is an important figure for an analysis of plan financing. From the angle of the source of finance, plan revenue component will be comprised of additional taxation, plan grant and balance from current revenues. We have the agreed additional taxation figures in the Third Plan document. The Third Finance Commission recommended certain figures of plan assistance to cover 75% of the plan revenue gap. (p. 33, Report of the Finance Commission, 1961). From these figures notional plan grants have been worked out. Though many states estimated a positive balance from current revenues at the time of the formulation of the plan in 1960, the Finance Commission in 1961 allowed for such positive balance in its estimates only for Maharashtra and Uttar Pradesh. The figures for these states under 'plans grants' have been marked * since the figures are not fully plan grants as recommended by the Third Finance Commission.]

TABLE 4.6

Additional Taxation Targets of States as a Percentage of State Tax Revenue

(Rs. Crores)

State	Addl. Taxation in III Plan (annual average)	Total state tax revenue in 1960–61	Percentage
1. Andhra Pradesh	10·6	83·6	13
2. Assam	3·2	32·7	10
3. Bihar	10·00	78·6	13
4. Gujarat	5·8	52·0	11
5. Jammu and Kashmir	1·6	15·0	11
6. Kerala	4·6	44·2	10
7. Madhya Pradesh	9·6	68·3	14
8. Madras	9·0	91·3	10
9. Maharashtra	10·4	115·4	9
10. Mysore	8·4	62·1	14
11. Orissa	4·6	34·5	13
12. Punjab	8·0	61·2	13
13. Rajasthan	6·4	43·2	15
14. Uttar Pradesh	21·8	135·1	16
15. West Bengal	8·0	94·8	8
All States	122·0	1,011·8	12

[Source: Third Five Year Plan and Report of the Finance Commission, 1961.]

because of lack of uniform application or otherwise and may not therefore include them in the plans. Similarly, for want of extra assistance, state governments may not be willing to take on schemes found necessary during the implementation of the plan, once the plan under the centrally assisted schemes is finalized. Such schemes can also be sent down in centrally sponsored bottles. Empirically, the only definition that is possible is that the centrally sponsored schemes are those assistance for which is given over and above the assistance assured for the state plan as a whole.

The state plan consists of the outlays on centrally assisted schemes and also the state's share of the outlay on centrally sponsored schemes. Under the centrally assisted schemes the patterns and procedures of assistance have been undergoing changes.[1] Assistance

[1]For a detailed description of the evolution of central assistance during the plans, see A. Premchand, *Control of Public Expenditure in India* (Allied Publishers), Chapter VI. See also Hanson, op. cit., Chapter X.

was first determined and given schemewise, and with no scope for adjustments of outlays, and rigid sanction procedures. During the first half of the second Plan, groups of schemes became units for adjustments and for determination of central assistance. Even after this there were operational difficulties and heads of development (i.e. a classified collection of groups of schemes) became units, but strangely enough, patterns for individual schemes have continued. The present position is that, broadly speaking, the state plan is broken down into a series of major heads of development containing various groups of centrally assisted schemes, some of which earn assistance and some of which do not. Loans and grants are allocated for each head of development, their quantum being arrived at mainly on the basis of schemes for which patterns are provided. With the exception of certain schemes like girls' education, control of communicable diseases and schemes under agriculture and co-operation where no diversion of outlay is possible, in respect of other schemes adjustments within a head of development are permissible. But diversions from pattern-schemes to non-pattern-schemes would require reference to the central ministry concerned. Reappropriations as between different heads of development will require consultation with the Planning Commission. Sanctions for centrally assisted schemes are issued by state governments without consulting the centre.[1] For centrally sponsored schemes costing over Rs. 25/- lakhs for the plan period or Rs. 10 lakhs in a year prior sanction of the central ministry is necessary.

The assistance for each year (within the ambit of the total plan assistance) is fixed by the centre for each state after discussion on the size of the annual plan when the outlay under each head of development is fixed. It is released on the basis of performance, individually for grants and loans under each head of development, the formula being

$$\frac{\textit{actual outlay}}{\textit{agreed outlay}} \times \text{loan/grant fixed for the head of development.}$$

There is a ceiling in central assistance for each head of development and for the plan as a whole. If a state does not fulfil the outlay estimated under a particular development it will lose proportionately from the fixed quantum of grant or loan. If a state exceeds its agreed outlay on a head of development it will not earn a higher quantum of grants under that head of development though if the

[1]Irrigation and power schemes are cleared with the Central Water and Power Commission.

overall state plan is fulfilled, it may get an equal amount of loan to make up the total plan assistance assured. Thus if the annual plan outlay as a whole is reached, the promised quantum of central assistance is available but the grant element would depend on whether the agreed outlay is fulfilled under each head of development which carries a grant. If the annual plan outlay as a whole is not reached, the central assistance given would be less than originally fixed, the exact quantum of grants and loans depending on the performance under each head of development.

What we have just now described above is the method of actual calculation of central assistance. The process has necessarily to be such that the entitlement is known only after the performance. But the states will not be able to incur heavy expenditures on plan schemes and wait for central assistance to be released later. Since 1958, the ways and means difficulties of the states have been borne in mind. The central assistance decided upon for a particular year is usually released in nine equal monthly instalments from the month of May of every financial year. The amounts are not finally debited to the heads of appropriation but are kept as 'ways and means advances'. The states send quarterly returns on the progress of expenditure under each scheme. Heavy shortfalls under them would mean a cut in the release. For the last quarter of the year they also send 'anticipated actuals' so that before the finanical year is out, sanction for the release for central assistance can be made. The grants and loans thus sanctioned are then reclassified from being ways and means advances to actual expenditure debitable under the relevant heads of appropriation. The process of course cannot stop here since the actual expenditure for a financial year will be known only some months after the close of the year. Once the actual expenditure is known the short payment or excess payment is worked out and further payment or refund, as the case may be, is made.

It will be seen from what has been said above that the procedure for central assistance is complicated and is liable to criticism. The procedure can affect the states in three ways: (1) central control (2) constriction on operational flexibility in the implementation of the plan and (3) delays due to centre-state consultations. Modifications in the procedures for adjustments and sanction of schemes have lessened the difficulties arising from the latter two items. As regards central control, we shall be taking it up as part of the nexus of centre-state relations in another chapter.

What has been the part played by state finances in the planning endeavour and what has been the effect of planning on state finances? The first question we have answered in some detail while discussing the resource-composition of the state plans and the second question may now be answered briefly. The revenues of state governments have decidedly benefited (particularly sales tax, motor vehicle tax and electricity duties) from the measures of economic development, though it is difficult to assess precisely the extent to which the state revenues have improved on this account. On the expenditure side it is well known that the phenomenal growth in expenditure of the state governments has been due to the Plans. Not only development expenditure but non-development expenditure too has grown due to planning, as it has rendered the growth of administrative and other services necessary because of urbanization etc. The emphasis on planning naturally leads to greater emphasis on the capital budget and the indebtedness of the states to the public and to the centre has meant heavy burden on interest charges with a consequent strain on the revenue account. This is another reason why non-development expenditure has developed a growth of its own, side by side to the emphasis on development expenditure. The magnitude of the impact of planning on state finances could be seen from the fact that for the Third Plan the expenditure on debt services for all the states (including appropriation for reduction or avoidance of debt) has been Rs. 934 crores[1] which is more than their Rs. 610 crores of additional taxation for the Third Plan. If one were to put it contentiously, the centre is giving grants from one hand for plan purposes and taking away part of them on the other by way of interest charges on debts, getting in the process the benefit of control that grant-in-aid means.

Planning has also affected the ways and means position of state governments. No doubt the revised procedure has enabled the states to draw ways and means advances for practically every month for the execution of plan schemes. Even so heavy plan expenditure has resulted in ways and means difficulties, exhaustion of cash reserves and securities and the emergence of unauthorized overdrafts with the Reserve Bank. As against these difficulties we have to consider the various benefits that the states have certainly derived from planning, particularly in matters of irrigation and power, education and public health.

[1]Reserve Bank of India Bulletin, May, 1966.

CHAPTER V

UNION-STATE RELATIONS—I

The subject of Union-State relations subsumes a variety of aspects in the political, economic, financial and administrative fields. We are concerned in this book only with the financial and planning aspects of Union-State relations. Even these aspects we shall discuss only in broad outlines as a detailed discussion in all aspects will require a book in itself.

As we have indicated in the first chapter, the field of Union-State relations has not been quiet. One hears many voices. First class constitutional issues have arisen. Every Finance Commission from the Second Finance Commission onwards has called for a review of the constitutional position without any result. The centre's attitude is dominated by the need for the fulfilment of the Plans and national objectives and by the need to keep erring states on the leash. As regards the states, they are satisfied with neither the quantum of resources transferred, nor their distribution *inter se* nor the controls that go with the transfer. It is true that in no federation have the central and state governments seen eye to eye. One should almost say that it would be undesirable if they did! But at present inter-state relations in India are not just spheres of friendly pugnacity but matters of considerable consequence and there are no easy solutions in sight. 'As things have developed financial relations are more complicated in India than in almost any other federation, certainly more than they are in the U.S.A., Canada, Australia or Nigeria. A formidable problem of coordination is implied'.[1]

It is not necessary for us to dilate on what is by now well known viz. the increasing magnitude of transfers from the centre to the states in all federations. Nor is it necessary to discuss the principles of central grant-in-aid as evolved in various federations. There is not much point in discussing them in the abstract and we will in any case come across some of these problems while discussing the Indian situation.

We may therefore proceed straightaway to an examination of the

[1]U. K. Hicks, 'Some Fundamental Problems of Federal Finance', p. 15, Supplement to *Capital* dated 23rd December, 1965.

'transfer problem' in India. Table 5.1 shows the quantum of transfer of resources from the centre to the states from the beginning of the First Five Year Plan and its relation to the total resources of the states.

TABLE 5.1

Transfer of Resources from Centre to States

(Rs. Crores)

Year	Resources made available by Centre				Expenditure of States			Percentage of	
	Shared Taxes	Grants	Loans	Total	Revenue Account	Capital Account	Total	2 + 3 to 6	5 to 8
(1)	2	3	4	5	6	7	8	9	10
1951–52	53	34	73	160	393	151	544	22	29
1952–53	74	36	112	222	417	155	572	26	39
1953–54	73	45	155	273	446	177	623	26	44
1954–55	72	51	221	344	494	206	700	25	49
1955–56	81	48	262	391	604	308	912	22	43
Total First Plan	353	215	822	1,390	2,354	997	3,352	24	41
1956-57	78	39	223	339	628	319	948	19	36
1957–58	121	114	284	519	684	347	1,031	34	50
1958–59	162	141	284	588	765	357	1,122	40	52
1959–60	170	182	295	647	870	396	1,266	40	51
1960–61	179	224	339	742	987	488	1,475	41	50
Total Second Plan	710	700	1,423	2,834	3,935	1,908	5,843	36	49
1961–62	178	217	452	847	1,121	509	1,630	35	52
1962–63	224	222	524	970	1,261	547	1,808	35	54
1963–64	260	231	624	1,115	1,412	690	2,102	35	53
1964–65	258	285	680	1,223	1,585	771	2,356	34	52
1965–66 (RE)	276	351	819	1,446	1,913	942	2,855	33	51
Total Third Plan	1,196	1,316	3,099	5,601	7,293	3,459	10,751	34	52
1966–67 (BE)	361	400	659	1,410	2,086	782	2,868	37	49

Source: Finances of State Governments, Reserve Bank of India Bulletin, May, 1966.]

The first thing strikes one is of course the sheer magnitude of central assistance. In fifteen years between 1951–52 and 1965–66, central assistance increased ninefold even though the overall resources of the state increased only over fivefold. Each Five Year Plan has witnessed an increasing quantum of central assistance, the second Plan doubling the first's and the third doubling the second's. The share of central assistance in the overall resources of the state rose from 29·4% in 1951–52 to 52·1% in 1964–65. From 1957–58 onwards the share has fluctuated from 50 to 54%.

It may well be asked how the states' autonomy can be said to be restricted when they are finding 50% of the resources themselves. Here it might as well be said that when it comes to it, it is not so much the percentage but the elbowroom in the margin that counts. We have seen already that the plan expenditure is largely financed by the centre. Whatever resources that the states may raise are already taken into account for the fulfilment of the Plan. The Finance Commission has also, by and large, only covered budgetary gaps.

The net result is that the States have very little scope for undertaking any new schemes of expenditure other than those included in the plan, except by cutting their non-plan expenditure or running a bigger deficit than they bargained for.

There are some changes in the components of central assistance which deserve to be taken note of but we must first enumerate the various items of central assistance. The item 'shared taxes' consists of

(1) Share of divisible taxes; (a) obligatorily shared viz. income tax; and (b) permissively shared viz. excise duties.

(2) Taxes and duties collected by the centre but wholly accruing to the states viz. (a) Estate duty and (b) additional excise duties.

This item, which is determined by the awards of the Finance Commission once in five years has shown an increase from plan to plan consistent with the buoyancy of the economy and the needs of the states. Within this group itself, income tax, the premier divisible tax, has exhibited a relative decline in importance because of (a) the shrinkage of the divisible pool consequent on the separation of taxation of company incomes from income taxation (in 1959); (b) powers of levy of surcharge of income tax by the centre outside the divisible pool; (c) emergence of excise duties as a major source of revenue; and (d) increasing needs of states.

As a matter of fact shared taxes as a whole have themselves been 'overtaken' in importance by grants. The absolute quantum of grants was less than that of shared taxes in the First Plan and was about equal in the Second. In the Third Plan it is decidedly more. Grants may be broadly classified as 'plan' and 'non-plan' grants but their nature and direction can be understood from the following exhaustive list over the three plans:

A. Grants met from revenue (of the Centre).

(1) Grants under Article 273.
(2) Grants under substantive portion of Article 275(1).
(3) Grants under Proviso to Article 275(1 .
(4) Grants under Article 278.
(5) Grants-in-aid under Section 74 of States Re-organization Act.
(6) Payments to police force.
(7) Grants for agricultural schemes, dairying, rural development etc.
(8) Grants for forest development.

(9) Grants for community development schemes.
(10) Grants for educational schemes, scientific departments and community listening schemes.
(11) Grants for medical and public health schemes.
(12) Grants for welfare of backward classes.
(13) Grants for small industries development schemes.
(14) Grants for relief and rehabilitation schemes.
(15) Grants for labour and employment schemes.
(16) Subsidy for acquisition of gold.
(17) Grants for natural calamities.
(18) Assistance for raising emoluments of low paid employees.
(19) Payments to state governments on account of reduction in their shares of Income Tax consequent upon changes in company taxation.
(20) Grants in lieu of railway passenger fares.
(21) Grants for development of border areas.
(22) Others.

B. Grants met from capital (of the Centre).

(1) Local Development works.
(2) Industrial housing and slum clearance.
(3) State roads of economic or inter-state importance.
(4) Grants from Central Road Fund.
(5) National water supply programme.
(6) Others.

Except for grants given on an *aa hoc* basis for unforeseen situations, non-plan grants are determined by the Finance Commission; Plan grants are determined by the Planning Commission. The increasing outlays on plans have necessarily meant substantial increase in plan grants.

But more than the grants it is the loans that have increased very substantially over the Plans. This is of course to be expected, since planned development involves large capital outlays. Open market borrowings by states have not been on the scale required for the capital needs of the states. Hence it is central assistance that has very largely financed the capital accounts of the states, as will be seen from Table 5.2

There are thus three broad categories of fiscal transfers viz. shared taxes, grants and loans and there are two bodies to determine the entitlements. 'Shared taxes' are the exclusive domain of

TABLE 5.2

Central Loans and Capital Accounts of States

(Rs. Crores)

Plan (1)	Loans from centre (2)	Capital Account of states (3)	Percentage of (2) to (3) (4)
First Plan	822	997	82%
Second Plan	1,423	1,908	75%
Third Plan	3,099	3,459	87%

[Source: Table 5.1]

the Finance Commission and 'loans' that of the Planning Commission.[1] 'Grants' is a common field, the Finance Commission deciding the non-plan side and the Planning Commission the plan side. The Planning Commission has no constitutional basis and the grants under Article 275 can be made only by the Finance Commission. Hence the Plan grants determined by the Planning Commission are given under Article 282 which is actually not a substantive provision intended for transfer of resources from the centre to the states but a 'miscellaneous' provision for grant-in-aid intended for validating expenditure outside the legislative power of the centre or the state.[2] But Article 282 has been overworked to a point where it has overshadowed its substantive brother, Article 275. The assistance given under Article 282 was 48·7% of the total in 1952–53 and went up to as much as 80·2% for 1961–62.[3]

Of course the plan and non-plan grants differ not only in respect of the authority determining their quantum and the Article of the Constitution justifying their existence, but in the basic fact that the non-plan grants under Article 275 are virtually unconditional whereas the plan grants are discretionary and conditional, varying

[1]Mr P. V. Rajamannar, Chairman of the Fourth Finance Commission and an ex-Chief Justice of Madras State considers that constitutionally the Finance Commission is not precluded from deciding loan assistance. *Vide* his minute to the Report of the Fourth Finance Commission, 1965, p. 89.

[2]p. 91 ibid.

[3]p. 40, Report of the Finance Commission, 1961.

from year to year and depending on the plan performance of the states. Plan grants, it has been considered, have to be of a flexible type, capable of revision according to performance and securing particular objectives. In the absence of any other provision relating to grants-in-aid, Article 282 has become the haven of discretionary plan grants.

Central assistance for the plans has thus proceeded on lines not anticipated by the framers of the Constitution. 'Chapter I of Part XII of the Constitution' writes Mr Santhanam, 'which regulates the financial relations between the Union and the States is based on two major assumptions. The first is that the main assistance required from the Centre would be in the nature of shares in taxes and grants towards the recurring revenue expenditures of the states. Though under Article 293, the Government of India is empowered to make loans to states or give guarantees in respect of loans raised by them, it was contemplated that normally the capital needs of a state would be met by its own borrowing. The second assumption is that the Finance Commission would be the chief instrument for determining the subventions and grants and the discretionary paragraph under Article 282 would be used only for special emergencies like famines or floods or other natural calamities. Both these assumptions have now broken down on account of the adoption of the policy of planned development under the guidance of a central Planning Commission.'[1]

The upshot of all this is that there is a controversial dualism in central assistance to states. The Finance Commission itself, instead of being the chief arbiter of the transfer problem, has become, by a process which it could protest against, but could not stop, an arithmetical agency for devolutions and a forum for ventilation of state grievances. To put it in the words of the Third Finance Commission, 'against this background, the role of the Finance Commission comes to be, at best, that of an agency to review the forecasts of revenue and expenditure submitted by the states and the acceptance of the revenue element of the Plan as indicated by the Planning Commission for determining the quantum of devolution and grants-in-aid to be made; and, at worst, its function is merely to undertake an arithmetical exercise of devolution, based on amounts of assistance for each state already settled by the Planning Commission, to be made under different heads on the basis of certain principles to be prescribed.'[2]

[1]K. Santhanam, *Transition in India* (Asia Publishing House, 1964), p. 116.
[2]Report of the Finance Commission, 1961, p. 35.

We shall devote the rest of this chapter to a consideration of the role of the Finance Commission in state finances. The other aspects of Union-State relations can conveniently be taken in the next chapter.

It is best to begin by looking at Article 280 which provides for the constitution of the Finance Commission:

'*Article 280:* (1) The President shall, within two years from the commencement of this Constitution and thereafter at the expiration of every fifth year or at such earlier time as the President considers necessary, by order constitute a Finance Commission which shall consist of a Chairman and four other members to be appointed by the President.

(2) Parliament may by law determine the qualifications which shall be requisite for appointment as members of the Commission and the manner in which they shall be selected.

(3) It shall be the duty of the Commission to make recommendations to the President as to—

(a) the distribution between the Union and the States of the net proceeds of taxes which are to be, or may be, divided between them under this chapter and the allocation between the States of the respective shares of such proceeds;

(b) the principles which should govern the grants-in-aid of the revenues of the States out of the Consolidated Fund of India;

(c) the continuance or modification of the terms of any agreement entered into by the Government of India with the Government of any State specified in Part B of the First Schedule under clause (i) of Article 278 or under article 306; and

(d) any other matter referred to the Commission by the President in the interests of sound finance.

(4) The Commission shall determine their procedure and shall have such powers in the performance of their functions as Parliament may by law confer on them.'

There is no doubt from a reading of this article that the Commission is intended to be the arbiter for transfer of resources from the Centre to the States. 'Under any other matters' the Commission could be asked to go into all relevant aspects of the finances of the state governments. The Finance Commission can thus occupy a pivotal role in the centre-state equilibrium of functions and finances.

The Commission itself is expected to be a highpower body and its composition is settled by parliamentary legislation. A law of Parliament provides that it should have a Chairman of certain qualifications; a member connected with Central Finances; an economist of repute; a public man who has been connected with the state finances; and a lawyer of the status of a High Court Judge.

The Commission is appointed once in five years. The period of its terms of reference did not coincide with that of the Plans for the first two Plan periods but the Third Finance Commission, appointed in 1961, was asked specifically to cover the period of four years commencing from April 1st, 1962 so that in future the period of recommendations of the Finance Commission would coincide exactly with the period of the Five Year Plans.

It is unnecessary for us to proceed on a lengthy disquisition on the recommendations of the four Finance Commissions appointed so far. For the sake of retaining the perspective, we may exclude all minor matters and refer to the general observations in the appropriate places in the book. For the rest, the Table 5.3 will summarize the recommendations of the four Finance Commissions.

Though the cardinal aim of transfer of resources is to meet needs and secure a large measure of uniformity in standards, the Finance Commissions have had to equate, in view of practical difficulties, budgetary needs with actual needs. The budgetary needs of the states are taken as the starting point for determining the assistance required, but these needs are adjusted with reference to certain other considerations. The budgets of the states are first of all reduced to a comparable basis by making adjustments for abnormal and non-recurring items of revenue and expenditure. Due allowance is besides to be made for 'clear cases' of failure of a state to maximize tax effort. Thirdly, in order not to place a premium on extravagance, a state's endeavour to secure reasonable economies in expenditure is reckoned with. Any state whose standards of social services are significantly lower than those of others should qualify for special assistance. Special disparities of states due to abnormal conditions beyond their control have to be provided for. Grants may also have to be made to certain states for the furtherance of objectives of national importance such as primary education, in respect of which they may be specially backward.

These were principles formulated by the First Finance Commission which its successors have accepted as unexceptionable. The Second Finance Commission also considered that the eligibility of a

TABLE 5.3

Main Recommendations of the Four Finance Commissions

(Rupees, Crores)

Assistance	(I) 1952–53 to 1956–57	(II) 1957–58 to 1961–62	(III) 1962–63 to 1965–66	(IV) 1966–67 to 1970–71
I *Income Tax*				
(a) Coverage	Includes taxes on company incomes, excludes surcharge	Includes taxes on company incomes,* excludes surcharge	Excludes taxes on company incomes and surcharge	Excludes taxes on company incomes and surcharge
(b) Share in divisible pool	55%	60%	66⅔%	75%
(c) Manner of allocation	80% population 20% collection	90% population 10% collection	80% population 20% collection	80% population 20% collection
(d) Quantum	278	375	555	**
II *Excise Duties*				
(a) Coverage	3 commodities	8 commodities	35 commodities	All commodities
(b) Share in divisible pool	40%	25%	20%	20%
(c) Manner of allocation	population	90% population 10% adjustments	Numbers of factors. Details not given	80% population 20% backwardness
(d) Quantum	46	153	394	**
III *Grants-in-aid* (revenue gap)				
(a) Coverage	Non-plan	Non-plan 11 states	Non-plan† 10 states	Non-plan 10 states
(b) Principles	Budgetary needs, tax effort, standard of services, special factors, etc.	Same	Same	Same
(c) Quantum	27	153	197	609
IV *Special Grants* (a) Coverage (b) Principles (c) Quantum	Nil	Nil	10 states Backwardness of communications 36	Nil
V *Additional Excise Duties* in lieu of sales tax				
(a) Coverage	—	5 Commodities	5 Commodities	5 Commodities
(b) Manner of allocation	—	Guaranteed amount on basis of erstwhile collections (32.54) Balance consumption-cum-population.	Guaranteed amount. Balance on percentage of increase in sales tax collections in states cum population.	Guaranteed amount. Balance share of each state in total sales tax collections cum population.
(c) Quantum	—	128	216	**

Note: Quantum is for each Plan period.

* In 1959, taxes on company incomes were classified as corporation tax.

**Depends on future collections.

† The Commission recommended grants for covering 75% of the plan revenue component but this was not accepted by the Government of India.

[Source: 1. Reports of Finance Commissions.
2. Reserve Bank of India Bulletins.]

state to grants-in-aid and the amount of such aid should depend upon its fiscal need in a comprehensive sense i.e. including the provision needed for the Plan. It also suggested that the gap between the ordinary revenue of the state and its normal inescapable expenditure should as far as possible be met by sharing the taxes. Grants-in-aid, it considered, should be largely a residuary form of assistance given in the form of generally unconditional grants. Grants for broad purposes may also be given but where those purposes are provided for in a comprehensive plan there will be no scope for such grants.

The same principles have been followed by subsequent Finance Commissions but all of them have been uniformly handicapped by paucity of data as well as semantic difficulties. 'Clear cases' of inadequate taxation were difficult to identify and the Finance Commissions had to assume that if a state raised the additional revenue which it promised for the Plan it should be reckoned to have done its part. The Third Finance Commission reported that it had, like its predecessors, been compelled 'to cover the budgetary gaps of all states whether caused by normal growth of expenditure the maintenance costs of completed schemes and mounting interest charges, or even by a measure of improvidence'.[1]

It would be beyond the scheme of this book to have any detailed discussion regarding the possibility of evolving indices for judging the tax efforts of states and their efficiency in expenditure.[2] The Finance Commissions have not found it possible in practice to give any great weightage to these aspects. Nor is the approach of the Finance Commissions to more or less identify actual needs with budgetary deficits free from criticism. It is certainly not an ideal touchstone for transfer of resources in a federation. 'The device of balancing current budgets on request' says Mrs U. K. Hicks 'is if anything worse than the tax targets sanctioned. It gives a state even less incentive for good management. This method has been followed precisely by the British Treasury in the West Indies and its failure has been most striking. The more ambitious islands widened their deficits by tucking into current expenditure a number of items which were strictly speaking on capital account. The lazy ones stopped bothering to collect taxes. One island after another achieved a deficit and nestled comfortably into the fold of grant-

[1]p. 38, Report of the Finance Commission, 1961.

[2]See, however, K. V. S. Sastri, *Federal-State Fiscal Relations in India,* (Oxford University Press, 1966).

aid.'[1] It cannot be said however that in India the approach of covering budgetary deficits needs to be termed as a 'failure', though some states have not been free from the temptation to plan for as large a deficit as possible in a pre-Finance Commission year.[2]

In the context of general devolutions and differential grants-in-aid, the principles of allocation for devolution rather lose their importance. The Finance Commissions, it would have been noticed, have largely relied on population and to some extent on derivation and they have not been able to adopt any other criteria in the absence of adequate statistics. It has been shown that population need not necessarily have redistributional effects if the richer states are also more populous. Besides density of population can also have an important bearing on needs. The state income, it has been suggested, would be a better guide for such allocation.[3] The ratio of the quantum of resources transferred to the non-plan revenue account of the state may also provide an index of the equity of the transfers. But the precise adjustments that the Finance Commissions make to ensure equity are not (rightly) published, with the result it is very difficult to say accurately how far the awards of Finance Commissions, given their approach, are fair to the various states. Hence ultimately, it is not some magical formula of central assistance but the composition of the Finance Commissions and their standing that have, as in the case of the Common-

[1]U. K. Hicks, 'Some Fundamental Problems of Federal Finance' in Supplement to *Capital* dated 23rd December, 1965, p. 11.

[2]Grants-in-aid seem to be never unmixed blessings. In Australia Mr H. P. Brown listed as the most important disadvantages of the grants system the fact 'that the state financial position is now so hopeless, that the state governments have little incentive to set their own house in order by such actions as raising railway rates and other charges or by exercising a reasonable restriction on activities which are not essential. Each state is constrained to make an annual approach to the Commonwealth Government for an ex-gratia increased grant and there must be a growing feeling that this implies a Commonwealth responsibility for the ultimate solvency of the states'. (H. P. Brown, 'Some Aspects of Federal-State Financial Relations,' in Geoffrey Sauer (ed.) *Federalism, an Australian Jubilee Study* (Melbourne, 1952). Prof. Prest mentions the same feature as the greatest weakness of the grants system (Wilfred Prest, 'Economics of Federal-State Finances'). Both quoted in *Public Expenditure in Australia*, B. U. Ratchford. (Duke University Press, Durham, N.C.)

[3]See U. K. Hicks, *ibid.* She has shown how striking the differences are between allocations of income tax and excises according to the Third Finance Commission and what such allocations would be if based on state incomes. But this does not provide us with a complete picture, as grants-in-aid are not taken into account.

wealth Grants Commission in Australia,[1] ensured acceptance of their awards by states without any significant protest.

The pity is that even if the Finance Commissions are possessed of a magical formula they can only produce half a rabbit out of the hat. This is due to their sphere of action being restricted, by enforced self-denial, to the non-plan revenue account. The Second Finance Commission made a reference to this growing anomaly arising out of the overlapping jurisdiction of the Planning Commission and the Finance Commission and the desirability of eliminating the necessity of making two separate assessments of the needs of the states. The Third Finance Commission came out much more openly but its brave sortie into the plan camp was repulsed. Keeping in mind the difficulties arising from the overlapping jurisdictions of the Planning and Finance Commissions and the various representations made by the states before it, it recommended differential grants-in-aid to enable the states, along with any surplus out of the devolutions, to cover 75% of the revenue component of their plans. The balance of 25% would be released by the Planning Commission after annual review. 'Our purpose in making these suggestions and recommendations', it said 'is twofold; first to secure the observance of the priorities of the Plan in regard to programmes of national importance, and secondly, to encourage and enable the State Governments to plan their affairs on a sounder and more realistic financial base and to discourage demoralization which dependence inevitably breeds'.[2] The safeguard in the utilization of this assistance for the purpose was, in its view, provided by Article 275 of the Constitution.

The competence of the Commission to make this recommendation was not in doubt. The sanction was, of course, Article 275 of the Constitution.[3] The Government of India, however, did not accept this recommendation of the Commission not because it thought the Commission had travelled beyond its sphere but because of practical considerations. It considered that it was neither necessary nor desirable to accept the recommendation since there would be no real advantage in the states receiving assistance for their plans,

[1]cf. A. H. Birch in *Federalism and Economic Growth*, ed. U. K. Hicks, (Allen and Unwin). For a comparative study, see B. N. Ganguli, 'Federal-State Financial Relations in U.S.A., Canada and Australia and their lesson for India' in N. V. Sovani and V. M. Dandekar ed., *Changing India* (Asia, 1961).

[2]p. 32, Report of the Finance Commission, 1961. See also *Federalism in India* by Asok Chanda (Chairman of this Commission), (George Allen and Unwin), 1965. Chapters VI and VII of his work deal with the subject of this chapter.

[3]p. 89, Report of the Finance Commission, 1965.

partly by way of statutory grants-in-aid by the Finance Commission and partly on the basis of annual reviews by the Planning Commission.[1]

The Government of India was obviously more in accord with the Member-Secretary of the Commission, Mr G. R. Kamat, a senior civil servant[2] who had dissented with this recommendation. In his minute of dissent he argued that the states would not derive any major advantage from the majority proposal. 'It certainly does not add to their resources, nor does it put them in a position of greater autonomy than at present'.[3] His main objection to united and unconditional grants for Plan purposes was that they would weaken the machinery which now enabled the centre and the states to effectively coordinate the formulation and implementation of the Plans. There would be difficulties in achieving the nationally accepted targets in the more important fields of development, if it was possible for the states to divert funds to schemes of lesser priority. The Plan itself was not rigid and the resources for the Plan flowed from various sources. When all other components of the Plan which were closely connected were subject to review and variation from time to time, it was unwise to introduce statutory rigidity in central assistance alone. With an assured amount of central grants for the Plan went the risk of some states slackening in their tax efforts or postponing it.

We shall not at this stage go into the merits of either argument but continue the story with what the Fourth Finance Commission did. It adopted a different approach to this thorny problem. In paragraph 16 of its report it said: 'The Constitution does not make any distinction between plan and non-plan expenditure and it is not unconstitutional for the Finance Commission to go into the whole question of the total revenue expenditure of the States. It has been pointed out to us that the reference to 'Capital and recurring sums' in the first proviso to article 275(1) of the Constitution suggests that even capital expenditure need not necessarily be outside the scope of the Finance Commission. It is, however, necessary to note that the importance of planned economic development is so great and its implementation so essential that there should not be any division of responsibility in regard to any element of plan expenditure. The Planning Commission has been specially constituted for advising the Government of India and the state governments in

[1]p. 89, Report of the Finance Commission, 1965.
[2]Currently (1966), Secretary to the Planning Commission.
[3]p. 54, Report of the Finance Commission, 1961.

this regard. It would not be appropriate for the Finance Commission to take upon itself the task of dealing with the States' new plan expenditure'.

'We have not, however, taken the view' it said 'that the function of the Finance Commission is simply to recommend such devolution and grants-in-aid as would merely fill up the non-plan revenue deficit as reported by the states because such an approach will be extremely mechanical. We have reassessed the states' estimates in the manner detailed in a subsequent chapter. We have not taken budgetary deficits as a criterion for distribution in the case of divisible taxes and duties'.

Though the Commission this gave a temporary quietus to the controversy, the constitutional difficulties still remain. This has been clearly recognized by Dr P. V. Rajamannar, the Chairman of the Commission and an ex-Chief Justice of the Madras High Court, in a minute appended by him to the report. After analysing the constitutional aspects of Union-State relations, he came to the conclusion that the time was ripe to have a review of the Union-State financial relationship, particularly in view of the setting up of the Planning Commission. This review, he said, should be made by a special Commission which can approach the several problems that have arisen in the past and that are likely to arise in future objectively and realistically.

The main question is of course how exactly the Constitution should be amended. Should the constitutional provision provide for a dual scheme of transfer of resources, as Dr Rajamannar seems to imply? That the state governments have a tendency, like the income tax-evader, to show different estimates to the Finance and Planning Commissions is a difficulty which can be solved by a little coordination. But is it desirable in the long run to have two separate assessments by two bodies, however eminent they may be and however desirable it may be in the short run to have flexibility? Though differing in composition and approach, they cannot be said to have any objective different from the balanced development of the states. So the discussion whether there is or should be one section or two sections for channeling central assistance does not reach the heart of the problem which is really whether there should be one body or two bodies which may determine central assistance through one article of the constitution or two.

Before we turn to a discussion of the possibilities and suggestions, we may refer to a slightly different question viz. the need for continuity in the Finance Commission's work. A permanent Commission will not necessarily be the best solution for continuity, because permanence may create vested interests and a self-righteous rigidity. The need for a permanent study cell to assist the Commission with various studies and statistics has been emphasized by all the Commissions but the implementation of this recommendation has been honoured in form and breached in observance. On the recommendation of the First Finance Commission, a cell was established in the President's Secretariat but was later transferred to the Finance Ministry on the recommendation of the Taxation Enquiry Commission. The Fourth Finance Commission found that this cell consisted only of some ministerial staff. A senior officer with adequate research staff will have to be appointed if an intelligent and purposeful approach to state finances is to be made. A convention can be established by which such an officer is ultimately appointed the Member-Secretary of the next Finance Commission.

We may now take up the substantive question viz. how to avoid overlapping jurisdictions and double views on central assistance—a question on which many views proceed on the premise that the question is really one of what to do with the body. Though the Third Finance Commission wanted an independent Highpower Commission to go into the question, it suggested, quite cryptically, two alternatives. The first was to enlarge the functions of the Finance Commission to embrace the total financial assistance be afforded to the states whether by way of loans or devolution or revenues, to enable them both to balance their normal budgets and to fulfil the prescribed targets of the Plans. 'This would, we consider, be in harmony with the spirit and even express provisions of our Constitution. This would also make the Commission's recommendations more realistic as they would take account of the inter-dependence of capital and revenue expenditure in a planned programme'.[1]

The second alternative suggested by it was to transform the Planning Commission into the Finance Commission at the 'appropriate time'.

The Fourth Finance Commission probably decided that silence on this point was golden. Dr Rajamannar however emphasized two points which impinge on the issue at hand. One was that the Planning Commission should be given the status of an independent permanent statutory body. The other was that in order to remove uncertainty a definitive allocation by way of the percentage of the shares of the Union and states respectively in income tax and excise may be fixed by the Constitution itself.

[1]p. 35–6, Report of the Finance Commission, 1961.

The other set of suggestions has been made by Mr Santhanam, the Chairman of the Second Finance Commission, 'There is no scope for any increase in the states' share of the income tax. There is some scope for increasing the share of excise duty. But it does not require a Finance Commission which can after all only increase the percentage in a more or less arbitary manner. I would suggest that 75% of the proceeds of income tax and 50% of the proceeds of all excise duties should be distributed to the states on the basis of their population. After such distribution there should be no grant either under 275(1) or under 282 to the states generally. But for the very poor states a fixed percentage of central revenues, say, 5% may be set apart. The distribution of this account as well as a general review of the finances of the states may be entrusted to the Finance Commission if it is considered that changing the Constitution is not desirable. I would however prefer that article 280 should be repealed and articles 270 and 272 should be amended incorporating the percentages of income tax and excise duties and the method of sharing them. In that case the distribution of the grant to the poor states may be done on the recommendation of the Planning Commission at the beginning of every Five Year Plan.'[1]

Professor Hanson finds the ruminations of the Third Finance Commission contradictory and impossible of rational explanation.[2] He argues that if national planning is to be taken seriously, there can be no insistence on the separate existence of the Finance Commission. The functions of the Finance Commission are, however, not obsolete, but they, being inseparable from the process of planning, ought to be confided to the Planning Commission. Hanson's arguments are however based solely on the need to cut out the confusions of dualism in the context of planning. He has not examined the constitutional implications of his proposal which involves, particularly, giving a constitutional status to the Planning Commission. Constitutional issues are after all important and even if a satisfactory constitutional modus vivendi is evolved, it is not an issue on which universal appreciation of the position will make the path of constitutional amendment easy. Moreover, with the emergence of non-Congress Governments in a number of states, the Finance Commission gains a special relevance and is rescued from euthanasia.

We may now sum up the various possibilities of the situation: (1) To let the Finance Commission determine plan as well as non-

[1]K. Santhanam, 'Federal Financial Relations in India', in *Indian Finance*, 12th November, 1966, p. 785.

[2]See Hanson, *op. cit.*, Chapter IX, pp. 337–347.

plan revenue grants and loans as well; (2) To make the Planning Commission work as Finance Commission; (3) To enlarge a constitutionally recognized Planning Commission and to have separate constitutionally recognized planning and finance wings; (4) To formally abolish the Finance Commission and write definitive allocations of taxes and grants-in-aid into the Constitution.

The situation is perhaps not that desperate. If we keep the essentials in mind and look at the problem objectively, non-radical solutions are possible. To abolish the Finance Commission would be a counsel of despair. To write allocations into the Constitution will produce a degree of rigidity that will have to be experienced to be realized. What will happen if the Centre alters the definition of income tax or enlarges the divisible pool of taxes? Is population a good enough criterion? Is it possible to entrench some other criteria and if so in what manner? The criteria for plan grants cannot be written in except in very general terms. If planning has to be realistic, the composition of the Planning Commission and its approach have to be flexible. The Planning Commission and its work cannot be written into the Constitution except in general terms. The fact is that the Finance Commission and the Planning Commission each have a kind of role which the other cannot play. Any kind of merger or fusion can be only in form. The roles have to be different. The Planning Commission can confine itself to planning overall magnitudes and concentrating on physical programmes and performance, leaving it to the Finance Commission to decide the methods of central assistance to state plans. The views of the Planning Commission can be placed before the Finance Commission which will no doubt give them the greatest weight. There will have to be a common estimate of resources and a convention that the Planning Commission's estimate of the desired size of state plans will be the unquestioned starting point of the exercise. As an additional factor, a member of the Planning Commission can also be appointed member of the Finance Commission. The Finance Commission will in its turn keep in mind the flexibility needed in planning and also the need for effectively securing the fulfilment of national priorities. Its grants can be in two parts, one unconditional and the other conditional but not unduly conditional and subject to annual variation according to performance. Article 275 may be amended if necessary for this purpose.

How can grants be conditional but not unduly conditional? Here we enter another area of controversy with which we may begin the next chapter.

CHAPTER VI

UNION-STATE RELATIONS—II

Central assistance for plan schemes constitutes another area of debate in the realm of Union-State relations. We have already seen broadly how this assistance is determined and given.[1] Our task in this chapter is therefore only to look at it as part of the effect of the planning process in central-state relations. So far as central assistance is concerned it should be admitted that the approach of the centre has been based on trial and error and the centre has shown itself to be not incapable of resilience and change. But, as the states are apt to ask, should it take three five year plans to bring the mechanics of central assistance to a state where it is still unsatisfactory? Is the ritual of trial necessary where there is a patent error?

As we have already seen, central assistance comprises of centrally assisted schemes as well as centrally sponsored schemes. We shall concern ourselves here primarily with centrally assisted schemes. There are various aspects of the procedure over which the states can voice their grievances. One is over the quantum of plan assistance itself or of its break-up into grant and loan. We have already discussed the manner of determination of central assistance. But in the total assistance the grant-loan break-up used to be a hush-hush matter. States will naturally prefer more grants in the central assistance bag. Grants are prima facie determined with reference to outlays on schemes and the pattern of assistance for each scheme but they invariably undergo a (unjustified) cut from the Finance Ministry—a case of giving the cake and eating it too. The ultimate grant-loan break-up, therefore, does have an element of mystery. But, if there is any central arbitrariness in this regard, each state will have to take it up with reference to its individual position. It is not likely that the states may find identical points to argue collectively.

Another area of grievance can be the delays and bottlenecks in the flow of funds. But a good deal of improvement in the release of assistance has been effected since 1958 and the states cannot have much to complain about ways and means difficulties created by

[1]Chapter IV. There will be a slight repetition of central aid procedures in this chapter but that is to make the discussion self-contained.

procedural shortcomings. Likewise, in matters of sanction, things have progressed farther from the days when the states had to go to the centre for sanction for each and every scheme.

Constriction on operational flexibility has been another source of difficulty in the past and is still an area in which improvements are needed. The first step in this regard was taken during the Second Plan when reappropriation by states without prior consultation was permitted within groups of schemes. This was later on permitted for heads of development, which were a collection of groups of schemes. There are however quite a few exceptions to the freedom of reappropriation. No diversions are possible, for instance, from the proposed outlays on girls' education and communicable diseases. Diversions are also not possible from programmes of agriculture and cooperation. Central assistance for village and small industries is calculated for each group under that head and not for the head of development as a whole.

Subject to these exceptions reappropriations are possible between heads of development. These adjustments will however require the concurrence of the Planning Commission or the Central Ministries concerned. If a State Government diverts funds from one head to another without the necessary approval of the centre, it forfeits proportionately its rights to the grant admissible under the former, but if it achieves the state Plan target as a whole it earns the assistance so forfeited in the form of a loan called the 'Miscellaneous Development Loan'. To this extent, a state has considerable freedom to reappropriate so long as it fulfils the overall targeted expenditure and is prepared to bear the penalty of receiving a portion of assistance as loan rather than grant.

Patterns, it is therefore urged, have lost much of their meaning. Neither the quantum is fixed initially with reference to patterns, nor is the payment made ultimately with reference to them. They cannot even ensure that a particular agreed outlay will be spent in that form. The centre can very well ask, by the same token, why the states should raise a hue and cry about patterns, if they are not effective and the grants in a sense become unconditional. But the states are hardly likely to agree that patterns are the result of some benign oddity of the central government. For one thing, patterns have become a maze,[1] too numerous and complicated. There are over 200 patterns and some of them are for trivial matters on which no national objective is involved. It looks as though the essential

[1]A phrase used in the Memorandum of the Madras Government to the Fourth Finance Commission.

fact is forgotten that patterns are meant for selective promotion of important national objectives rather than for merely covering each and every rupee of central assistance. It is not as if that the gift of central assistance should in every case be wrapped and tied with a pattern.

What is more, the patterns create a situation, it is alleged, where the states have a tendency to choose schemes not for their intrinsic worth but because they carry a pattern of assistance from the centre. No state can afford to overlook such schemes for fear of being accused of having looked a gift horse in the mouth. This undermines the sense of financial responsibility of the states and creates a situation where their primary concern is to get the maximum from the centre rather than choose locally suitable schemes.

It will be pertinent to intervene at this moment on behalf of the centre to ask whether there are really instances where states produced good schemes and suggested patterns for them but the centre rejected them. To the author's knowledge, there is no information that any worthwhile scheme suggested by the state was arbitrarily rejected by the centre as not being fit for inclusion in the Plan. The real difficulty seems to be different and in a sense, basic. The deliberations of the state working groups on the formulation of the plan, follow closely the trends in deliberations of the central working group and if they draw any plan on their own it may be eventually unrealistic from the point of view of central assistance. No doubt the central working group itself contains the representatives of the states but unless such representatives effectively combine or there is an outstanding and vigorous exponent of a state's viewpoint, the central representatives will have their way on what schemes should be included in a particular head of development and what pattern of assistance, if any, they should carry. Though there is a theoretical possibility for an alert state to free itself of such shackles, the practical processes are such that the states seem to acquire a degree of mental dependence on the centre. It is this mental dependence that is more regrettable than the financial dependence. If the central ministries are keen on proliferating patterns, a climate is created where planning from below according to local needs becomes impossible of achievement indeed.[1]

People who argue that central assistance has clouded the judgment of states will quote the instance of the expensive staff setup

[1]According to Mr Santhanam, there is a 'vertical integration' of the centre and the states. About the deliberations of the Working Groups see Hanson, *op. cit.*, ch. X.

under the Community Development Programme which they initially undertook in view of the pattern of assistance. It is only some states that have later emancipated themselves from the central community development pattern and evolved setups on their own. Another instance quoted is the high scales of pay for teachers in engineering colleges which the states adopted at the instance of the centre though such scales were much higher than the general level of emoluments in the states.

Protagonists for the centre will refer to different instances. They will argue that the administration of the plans is after all with the states.[1] In important fields of legislation in which the centre was interested, such as land reforms and panchayati raj, each state has gone in its own way. In respect of land reforms the states first went faster than the centre wanted them to, though when it came to the imposition of ceilings on holdings of land, the reform was not quick and uniform. In matters like increasing water rates etc. to make the irrigation projects pay, the Planning Commission's advice has not been heeded by the states. They will also argue that whatever progress the states have achieved would have been impossible without central assistance. Administration and the sphere of state action is basically intact and the structure still federal.[2] The states have so much to do with plan administration that Paul Appleby thought that the centre was in fact only a large staff agency and the states had far more money and personnel than in any other federation.[3]

All this may be conceded but does not alter the fact that the rigours of central assistance are greater than is necessary. The centre itself has been forced to recognize the defects of a rigid procedure and the few instalments of simplification and rationalization have stemmed from this fact. The Planning Commission has stated that the present restrictions and manipulations of grants-in-aid have not been really effective in giving a proper lead to state governments in implementing state plans. While the Commission felt that the distribution of subjects under the Constitution between the Union and State governments does not give to the centre sufficient powers to ensure economic and social development of the country as a

[1]'The working of the system also revealed that there is little or no control (as distinguished from review) by the central Government'. A. Premchand, *Control of Public Expenditure in India* (Allied Publishers), 1963, p. 208.

[2]See S. P. Jagota, 'Some Constitutional Aspects of Planning' in *Administration and Economic Development in India*, ed. B. R. Braibanti and J. J. Spengler (Cambridge University Press, 1963).

[3]Report on Public Administration in India.

whole either on an adequate scale or on uniform lines, the Commission admitted that rigid procedures for administering central assistance inhibit action in the states.[1]

While there has thus been recognition of the inhibitory effect of central procedures, the Centre has been slow to give them up. This is partly because it feels the states are apt to err in their ways if they are given too much freedom. Such a feeling is not entirely divorced from experience but that is not the foundation on which the edifice of development can be built. More importantly, the argument is that national priorities cannot be ensured without conditional grants. The unlimited field of social development is a tempting pasture. Besides, the states are not likely to attach the same priority to certain schemes which are not so urgent to it as the centre may, viewing it from a national angle. We have already quoted the argument that a state surplus in food production will not have that much urgency as the nation as a whole has to increase agricultural production.

Any pattern of central assistance for economic development, has to strike an acceptable balance between the views of the centre and that of the states if it is to be practical and effective. For a change, the centre could take up a bolder programme of central assistance. It should shift the emphasis from mere financial control to physical evaluation and purposeful assistance. If the states fail to live up to the expectations of the centre it can revert to its more stringent methods, a fortiorari. A bolder scheme is particularly warranted at this stage of development when the centre as well as the states have had some experience in cooperative development and a measure of understanding of each other. Sufficient progress has been made in the coverage of the population on matters of national concern like primary education and health. Further programmes in this regard can only be qualitative and can be left to the discretion and individual conditions of states. The main fields in which central direction will be necessary are agriculture, irrigation and power. There will be other individual schemes like girls' education, family planning etc. where insurance against failure is necessary. Leaving aside capital expenditure, all revenue grants could be in the form of block grants except for 'essential schemes' where if the agreed outlays are not fulfilled, cuts may be made. For the rest, grants may be released proportionate to the total plan revenue expenditure. 'Essential schemes' should not of course

[1]Planning Commission's note prepared for a meeting of the National Development Council. *The Hindu*, 29th October, 1964.

be multiplied but kept within a limit of say, ten. They may have their own specific patterns, and central assistance for them may be released individually as this would also afford a much better opportunity than is available now for a scrutiny of the achievements under these schemes. The virtue of the proposal is that it gives a large field for block grants, but its success depends on keeping the list of essential schemes to a limited number. There is no reason why specific central interest cannot be confined to this number of schemes. It would however be noticed that loan expenditure is kept separate. There can be a parallel distinction between essential schemes and other schemes on the loans side also.[1]

What we have to say on the loans side has to be taken in conjunction with our proposal in another chapter for a composite scheme of borrowing by states. We have stressed more than once that a large amount of central financial control is also through loans. In order to alleviate the vertigo of central assistance, we are proposing separately that capital schemes that can be expected to pay their way should be financed by loans from a Central Loan Council and other loans may be provided by the centre or through open market borrowing. This would reduce excessive central control and improve financial discipline.

It may be argued that the procedure we are proposing is also complicated. But we must remember that a measure of complication cannot be avoided if conflicting objectives are to be secured. We are far from outlining this proposal as the only possible solution to the problem at hand but are only illustrating how solutions are possible for reconciling conflicting interests in the context of the overall interest of national development. We cannot but invoke at this juncture the observation of the Treasury Committee on Intergovernmental Fiscal Relations (1943), of the United States. 'Much valuable energy has been wasted unnecessarily in quarrelling over the proper spheres of the Federal Government and the States, when the seeds of solid achievement lie in the scantily tilled field of intergovernmental cooperation and coordination. Progress in this field requires some willingness to compromise, to surrender vested interests and to forget jealousies on the part of both the Federal Government and the States. The American governmental system

[1]This was written before the Fourth General Elections in 1967. With an array of non-Congress Governments whose financial policies are yet unsettled, it is not practical to expect the centre to move towards liberalization. But a more acceptable pattern of assistance will surely have to be thought of before the Fifth Plan. See also Chapter XIV.

has not been viewed as a unit by most public officials, with loyalties evoked and encouraged for the entire system'.[1]

We may now turn to certain related aspects of Union-State financial relations which have come into prominence from time to time, to show the trends in federal financial relationship.

That there is overlapping in the sphere of excise taxation and sales taxation is well known. Equally well recognized is the fact that the solution is not to give up sales taxation altogether. This was stressed by the Taxation Enquiry Commission as well as the Fourth Finance Commission which had both occasion to examine the question. On the recommendation of the Taxation Enquiry Commission however the Central Sales Tax Act was passed which enabled the levy of a uniform rate of sales tax (at present 3%) on goods entering inter-state trade. In 1957, a decision was taken by the National Development Council that the states would give up their sales tax on mill-made textiles, sugar and tobacco and the centre would impose an additional excise on these commodities, the net proceeds being distributed among the states, subject to the existing income derived by each of the states being assured to it. The principles of distribution have been left to the Finance Commissions which have distributed a sum of Rs. 32·54 crores on the basis of the previous collections, the balance being distributed on principles evolved by each Commission.

The states have from time to time complained that the measure has not been operated by the centre in fairness to them. Specific representations were made before the Fourth Finance Commission. The main point of the criticism was that whereas over the period 1957–58 to 1965–66 the rates of basic duties of excise on some of the items brought within the scheme were raised and even special duties of excise were introduced, the rates of additional duties of excise have remained unchanged. If the substitution had not taken place, it is argued, the States would have had the opportunity of raising sales tax rates on these items and would have also benefited from the rise in prices, the sales tax being an ad valorem levy. It was further argued that over the past eight years sales tax revenues have shown a much higher rate of growth than the yield from the additional duties of excise and that if the scheme had not been introduced the rate of increase in sales tax revenues from these items would have been closer to the rate of sales tax revenue on other items.

As against this, the Union government seem to argue that over

[1]Quoted in Hansen and Perloff, *State and Local Finance in the National Economy* (W. W. Norton & Co.), p. 134.

the period 1958–59 to 1965–66 the yield from additional duties has increased by as much as 45%, the increase in the yield from basic duties of excise on these commodities during the same period (excluding the yield from special duties of excise which fall in a distinct category) being hardly 21%. The items covered under the items of additional duties of excise are essential consumer items, and it is not as if the states would have just gone on increasing the rates. Indeed, on items of comparable nature like matches, kerosene, coal and vegetable products, sales tax rates between 1958–59 and 1963–64 have remained either altogether unchanged or increased so very little. An important reason why the Union Government had not revised the additional duties on excise with every change in basic duties is that sugar and textile are the items in the case of which downward adjustments have often to be made and the Union Government did not want that the states' revenues should be adversely affected by these downward adjustments. It is only in the case of tobacco that basic duties have been increased and never lowered. The increase in the sales tax revenues in the States is, inter alia, due to enchancement of rates in the case of luxury and semi-luxury articles and coverage of new items. It was argued that it would not be too correct to assume that the States would have managed to realize the same rate of increase in the sales tax revenue from these items as they have realized in the case of the total sales tax yield.

The Fourth Finance Commission did not consider it necessary to go into the validity of the argument for or against the manner of implementation of the scheme by the Union Government. It however felt that if some sort of institutional arrangement existed and both Union and the state governments had the opportunity of explaining each other's views, the implementation of the scheme would have been considerably better and misunderstandings less. We shall be seeing a few more items where institutional arrangements would have produced better results.

Union-State financial relations come into prominence when the central government enhances the emoluments of its employees and the state governments are consequently forced to adopt similar increases in the state without the necessary wherewithal to finance such increase. The states have then to look to the centre for assistance. This process began during the Second Plan when the centre increased the emoluments of its employees and also gave assistance to the states for increasing the emoluments of low-paid employees. This assistance was given till the year 1960–61 since the further requirements were to be taken note of by the Finance Commission. Subsequently there have been periodical increases in dearness allowance by the central government. State governments have complained that the unilateral increases forced on them are difficult to finance since their resources on the non-plan and plan account have been fully taken into account by the Finance and Planning Commissions.

In 1959, the Centre passed a legislation which classified income tax paid by companies as corporation tax and thus excluded it from the divisible pool. The loss to the states by way of loss in the share of income tax was made good till 1960–61, after which the Finance Commission was expected to take charge of the matter. The states have considered this a case of unilateral decision by the Centre which has also affected the potentialities of the growth of income tax which is part of the divisible pool. A related complaint has been that the Centre has levied a surcharge on income tax which is outside the divisible pool. An increase of this surcharge rather than of the basic rates of income tax would avoid any share-out of the increase with the states. Thus the charges of central revenue being managed to the prejudice of state finances have been made from time to time.

Another significant matter in the context of Union-State relations is the tax on railway passenger fares. This tax was imposed in 1957 and the Second Finance Commission was requested to make recommendations as to the principles which should govern the distribution of the net proceeds under Article 269 of the Constitution. The Commission decided that the proceeds of the tax should be distributed among the states in the ratio of passenger earnings. However this tax was repealed in 1961 by merging it in the basic fares. The Union Government however decided, in pursuance of the recommendations of the Railway Convention Committee, 1960, to make an ad-hoc grant of Rs. 12·5 crores per annum to the states in lieu of the tax for a period of five years from 1961–62 to 1965–66. The Third Finance Commission recommended the same principles of distribution, namely passenger earnings. The Fourth Finance Commission also adopted the same principles and also the figure of Rs. 12·5 crores though at the time of this recommendation the Railway Convention Committee had not decided the future quantum of grant. The merger of the tax with the fares has been criticized by the state governments particularly on the ground that besides the impossibility of raising the rate which a tax would have provided, the fixation of the grant at the level of Rs. 12·5 crores has deprived them of the benefits of elasticity accruing from increased passenger earnings.

We have already given an instance of co-ordinated decision in respect of sales tax with reference to its overlapping effects on excise duties. An instance of co-ordination among states regarding standardization of the rates of sales tax at the behest of the Centre may also be given. At the suggestion of the Union Finance Minister all the States agreed to have a uniform single point tax, firstly at 7% and later at 10%, on certain luxury articles. But in the field of sales tax, problems of co-ordination among states and between the states and the centre are likely to arise from time to time. The Fourth Finance Commission was asked to make recommendations in regard to (1) the effect of the combined incidence of a state's sales tax and Union duties of excise on the production, consumption or export of commodities of products, the duties on which are sharable with the states and (2) the reductions, if any, to be made in the states' share of Union excise duties if the sales tax rates levied by the states exceed a certain specified ceiling. The Commission found that any conclusion regarding the incidence of sales tax and the union excise duties was not possible in the absence of the necessary studies and statistics. The Commission also considered that in the absence of such useful data, there would be no point in proceeding to the question of fixing a ceiling in respect of sales taxes.

The states of course protested against any such proposal very strongly and rightly too. As regards co-ordination in the matter of excise and sales tax the question of co-ordination arose only in the case of forty-nine items since the other items were either governed by the Central Sales Tax Act or by the decision to have a uniform levy of 10%. The States' view was that if on proper study of facts, in the case of selected items out of the forty-nine items, a co-ordinated tax policy was called for that could certainly be effected. But the proper course according to them for such co-ordination is not the imposition of a financial sanction in the form of a reduction in the share out of devolution items but a periodical exchange of views between the Union and the state governments on problems of taxation and related subjects with a view to evolving co-ordinated lines of action.

This brings us to the general question of inter-governmental co-ordination and the need of institutionalizing such co-ordination. From what we have said above there could be no doubt that such co-ordination is necessary, particularly in view of the feeling among the states that the temptation on the part of the Union government to neglect sharable taxes is inescapable. Contingencies of this type cannot however be said to have been not envisaged by the Constitution. Article 274 provides in effect that no proposal which in any way affects the existing or prospective financial interests of a state shall be presented to Parliament except on the recommendation of the President. Obviously what the constitution expected was that the procedure relating to the recommendation of the President would involve consultation with the states but in practice this does not seem to be the case. The Fourth Finance Commission said that though the procedural requirements of Article 274 were all along observed, such an observance might be capable of improvement in such manner as to more fully carry out the purpose of the Article and would command greater reassurance with the states.

An institution for inter-governmental co-ordination is therefore a desideraturm in the Union-State financial relations in India. The Finance Commission would no doubt be such an institution par excellence but meeting as it does once in five years, it may not be available for consultation and co-ordination at other times. The National Development Council[1] is a body which is quite representative and high powered to consider such questions authoritatively. But neither the Council nor its sub-committees have the time to go into specific financial matters in requisite detail. Hence the appropriate solution would be to make the Finance Commission Secretariat (which, we have suggested in the previous chapter, should be manned by a senior official who is or could become member-secretary of the Finance Commission) the co-ordinating body. Ad-hoc decisions arising in between two Finance Commissions could be co-ordinated and reached through the mechanism of this Secretariat. Once the Finance Commission is appointed it may as a matter of course go through all cases that had arisen in the past five years and suggest any changes if necessary.

There is also a great need for co-ordination among states themselves and for exchange and compilation of information on financial and taxation matters. Such an institution evolved on proper lines can be a valuable focal point for all matters connected with state finances and their relation to the centre. In the United States the Council of State Governments maintains in Chicago a sort of informal capitol of the states with subsidiaries, a magazine and a library. Originally supported by philanthrophic grants it now receives regular support by contribution from states. It has taken a leading

[1]It is a body whose membership includes the Prime Minister and Central Ministers, Chief Ministers of States and Members of Planning Commission. It has no constitutional basis.

interest in the federal coordination problem.[1] A similar body would be of great value in the present Indian context.

Such a body can be either a purely research body or a militant guardian of the interests of the states so far as combined action of the states as against the centre is required. The National Development Council can itself provide the forum for combined action by states if Chief Ministers could have the time to come to some kind of understanding among themselves. But considering the demands on its time, the National Development Council may not be in a position to have a patient discussion of the questions. In that case there will be no alternative but to have a separate Council of State Governments which can take this role. No doubt, it is ultimately the principle of cooperation and understanding that is more necessary in Union-State relations than mere institutions. But an institution as such is necessary in India in view of the paucity of information on state finances and the problems of consultation among states. Even if there is an institutional setup for consultation among the state governments it is doubtful whether the states will necessarily be able to arrive at a uniform stand on all matters as each state is likely to look at issues with reference to its own financial vantage.

Another suggestion for an inter-governmental institution can be for a Central Local Council as in Australia. But this we shall be discussing in the chapter relating to the debts of state governments.

[1]H. M. Groves, *Financing Government*, 1964, p. 452.

CHAPTER VII

TAX REVENUE

We shall now consider the tax revenues of the states. There are various aspects of the question of taxation by states and studies of public finance usually concern themselves more with taxation than with any other aspect. But in the scheme of our study, taxes alone cannot be dealt with at any disproportionate length. We shall confine ourselves to the trends in the magnitude and structure of state taxation and to the general problems involved in the exploitation of the sources of revenue. In the next chapter we shall attempt an assessment of the effort of additional taxation put forward by the states for the implementation of the plans and shall also try to identify the directions in which the revenues may be augmented.

We may begin by looking at the role of state taxes in the context of the total revenues of the states. Table 7.1 shows the total revenues of the states broken down into state taxes, devolutions, non-tax revenues and grants-in-aid.

The first thing that strikes one is that the revenues of the states as between 1951–52 and 1966–67 have grown by nearly five and a half times. Though the change in accounting classification in 1962–63 would inflate the figures from that year onwards, the fact remains that there has been a substantial growth of state revenues. The revenue for the year 1966–67 is far higher than the average for the Third Plan. State taxes which formed 58% of the total revenues in 1951–52 have gone down to 44% of the total in 1966–67 and this decline in importance has been more or less matched by the emergence of grants-in-aid from central government as an important factor. This increased from 6% in 1951–52 to 19% in 1966–67. Devolution or non-tax revenues have not gained or lost appreciably. State taxes themselves have shown considerable buoyancy due to the growth of the economy and also due to the widening and deepening of the tax structure. In terms of current prices, the state taxes in 1966–67 are four times that in 1951–52.

It would be interesting to compare at this stage itself the quantum of the state taxes and the volume of non-development expenditure as it would define the role of state taxes in the total financial activities

TABLE 7.1

State Taxes in Total Revenue of States

(Annual averages, Rs. Crores)

	1951–52	Percentage	Average in 1st Plan	Percentage	Average in 2nd Plan	Percentage	Average in 3rd Plan*	Percentage	1966–67**	Percentage
State Taxes	228	58	253	54	389	48	662	45	933	44
Devolutions	53	13	69	15	132	16	239	16	347	17
Non-tax revenue and miscellaneous	90	23	107	23	202	25	311	21	430	20
Grants in aid	25	6	38	8	85	11	251	18	402	19
TOTAL	396	100	468	100	808	100	1,463	100	2,112	100

*RE for 1964–65 and BE for 1965–66.

**Budget estimates

[Source: Reserve Bank of India Bulletin, May, 1966.]

of the states. Table 7.2 shows this position. It will be seen that the surplus of state taxes over non-development expenditure has been neither substantial nor has it grown as a percentage over the years.

TABLE 7.2

State Taxes and Non-Development Expenditure

(Rs. Crores)

	1951–52	Average I Plan	Average II Plan	Average * III Plan	1966–67**
State Taxes	228	253	389	662	933
Non-Development Expenditure	196	229	344	619	889
Surplus	32	24	45	43	44

Note: 1. *RE for 1964–65 and BE for 1965–66, ** BE for 1966–67.
2. Annual averages.

[Source: Reserve Bank of India Bulletin, May, 1966.]

Before we proceed to discuss the role of individual state taxes, we may have an idea of the expansion of state revenues in individual states. Table 7.3 shows the position from 1951–52 to 1955–56. Table 7.4 gives the position between 1957–58 and 1960–61 (omitting 1956–57 because of states' reorganization) and Table 7·5 gives the position from 1961–62 to 1966–67.

It will be seen that the percentage of tax revenue in the total revenue has varied from state to state. Excluding Jammu and Kashmir, Orissa and Mysore had a percentage of 29 and 30 only in 1957–58 as against Bombay and West Bengal which had a percentage of 57 and 53 respectively. The rate of growth between 1957–58 and 1966–67 has also varied from state to state.

We may now proceed to individual taxes in the portfolio of the states and their rate of growth over the years for the states as a whole. Table 7.6 gives the relevant figures. Among the major taxes, by far the highest growth has been from the general sales tax while taxes like motor vehicles tax and electricity duties have also grown in prominence. Table 7.7 will compare as between 1951–52 and 1966–67 the share of each of the taxes to the total state taxes.

TABLE 7.3

Growth of State Taxes in First Plan

(Rs. Crores)

State	1951–52 Amount	1951–52 Percentage to total revenue	1955–56 Amount	1955–56 Percentage to total revenue
1. Andhra Pradesh	..	..	14	59
2. Assam	6	51	10	46
3. Bihar	15	53	18	46
4. Bombay	34	56	48	58
5. Madhya Pradesh	12	52	8	48
6. Madras	36	61	12	43
7. Orissa	5	45	8	40
8. Punjab	8	46	10	42
9. Uttar Pradesh	27	53	42	52
10. West Bengal	23	61	28	56
11. Jammu and Kashmir	3	59	1	16
12. Hyderabad	20	68	17	63
13. Madhya Bharat	7	63	8	48
14. Mysore	6	45	8	40
15. Pepsu	4	70	5	51
16. Rajasthan	11	72	10	44
17. Saurashtra	3	39	5	37
18. Travancore-Cochin	9	49	9	47

[Source: *Report of the Finance Commission*, 1957.]

We may now discuss very briefly certain important taxes in the portfolio of the states. We may begin with land revenue which is the oldest of the state taxes. There have been considerable variations among states in the method of land revenue settlement as well as calculation of the tax due. The British system of land revenue administration went in terms of 'settlements' of land revenues which were done after detailed survey and classification.[1] In a few

[1]The different systems of settlements that have evolved in India can be broadly classified as: (i) permanent or those in which assessment was fixed in perpetuity and (ii) temporary or those in which assessment was fixed for a definite period. They can also be classified as: (i) zamindari, in which assessment was fixed on an estate held only by a landlord, (ii) mahalwari, in which assessment was fixed on a village or mahal jointly and severally on the whole village community and (iii) ryotwari, in which assessment was fixed on the holding of a ryot. See the Report of the Taxation Enquiry Commission, 1953–54, Vol. III, chapter I for a detailed description.

princely states the British system was adopted but in other princely states there were no regular systems of land revenue. States like Punjab, Maharashtra, Madras, Assam and Mysore belong to one category in which practically all the land had been surveyed and measured and settled on some definite principles. On the other hand in states like Rajasthan and portions of Gujarat and Madhya Pradesh there were large portions of unsurveyed and unsettled land. In between were the states like Uttar Pradesh and portions of Madhya Pradesh where there were regular settlements of lands but owing to the prevalent zamindari system several intermediaries had crept in. In West Bengal, Bihar and Orissa there was no organized land revenue system at the lower level because of permanent settlement.

TABLE 7.4

Growth of State Taxes in Second Plan

(Rs. Crores)

State	1957–58		1960–61	
	Amount	Percentage to total revenue	Amount	Percentage to total revenue
1. Andhra Pradesh	32	51	40	50
2. Assam	13	43	12	35
3. Bihar	20	40	31	39
4. Bombay	74	57	24*	43
5. Jammu and Kashmir	1	11	2	12
6. Kerala	13	47	20	45
7. Madhya Pradesh	21	41	27	38
8. Madras	32	51	42	45
9. Mysore	17	30	24	29
10. Orissa	6	29	9	23
11. Punjab	19	44	25	41
12. Rajasthan	15	48	18	41
13. Uttar Pradesh	50	48	57	39
14. West Bengal	36	53	48	51
15. Maharashtra	..	..	61	55

*Pertains to Gujarat.

[Source: *Report of the Finance Commission*, 1961.]

TABLE 7.5

Growth of State Taxes in Third Plan

(Rs. Crores)

State	1961–62		1965–66	
	Amount	Percentage to total revenue	Amount	Percentage to total revenue
1. Andhra Pradesh	42	43	65	42
2. Assam	15	41	25	35
3. Bihar	35	43	50	46
4. Gujarat	27	41	44	41
5. Jammu and Kashmir	2	9	3	11
6. Kerala	23	44	37	44
7. Madhya Pradesh	29	37	51	43
8. Madras	43	44	72	44
9. Maharashtra	67	56	111	52
10. Mysore	27	30	43	35
11. Orissa	10	22	19	23
12. Punjab	30	38	50	39
13. Rajasthan	21	46	34	43
14. Uttar Pradesh	59	36	83	33
15. West Bengal	53	52	77	51

Note: 1965–66 figures are Budget Estimates.

[Source: *Report of the Finance Commission*, 1965.]

The states have also adopted different methods in calculating the land revenue payable to government. The Taxation Enquiry Commission classified the methods in the following categories: (1) net assets or economic rent (2) net produce or annual value (3) empirical (4) rental value (5) capital value and (6) gross produce.

The implementation of land reforms mainly during the First Five Year Plan involved reassessment of land previously held through intermediaries and this added substantially to the land revenue in some states. The place of land revenue in the fiscal system of each state varies with reference to its past history as well as the nature of the lands and the irrigation facilities which determine the size of land revenue. Table 7.8 will bear this out.

It is well known that states have not been keen on substantial taxation of land. The Table 7.9 will show how slowly land revenue has increased in spite of increase in prices.

TABLE 7.6

Individual State Taxes

(Rs. Crores)

	1951–52	Average in I Plan	II Plan	III Plan	1966–67†
Land Revenue	48	65	91	112	121
Agricultural Income Tax	4	5	8	10	11
Stamps and Registration	26	27	37	60	74
Urban Immovable Property Tax	2	2	2	3	5
State Excise	49	46	48	74	99
General Sales Tax	54	62	116	243	371
Sales Tax on Motor Spirit	5	5	10	23	37
Motor Vehicles Tax	10	13	25	53	70
Entertainment Tax	6	6	10	21	30
Electricity Duties	4	5	10	26	44
Other taxes and duties	20	17	32*	37	60
TOTAL	228	253	389	662	922

*Includes tax on railway fares.

†Budget Estimate for 1966–67 without reckoning additional taxation.

[Source: Reserve Bank of India Bulletins.]

TABLE 7.7

Importance of Individual Taxes

Tax	As a percentage of state tax revenue in 1951–52	1966–67
Land Revenue	21	13
Agricultural Income Tax	2	1
Stamps and Registration	11	8
Urban Immovable Property Tax	1	1
State Excise	21	11
General Sales Tax	24	40
Sales Tax on Motor Spirit	2	4
Motor Vehicles Tax	4	8
Entertainment Tax	3	3
Electricity Duties	2	5
Other taxes and duties	9	6

[Source: Table 7.6.]

TABLE 7.8

Role of Land Revenue in the States

(Rs. Crores)

State	1951–52	1966–67†
1. Andhra Pradesh	..	15·01
2. Assam	1·81	5·15
3. Bihar	1·45	12·67
4. Gujarat	6·14*	6·06
5. Jammu and Kashmir	0·62	0·62
6. Kerala	..	1·63
7. Madhya Pradesh	4·39	9·67
8. Madras	6·57	6·33
9. Maharashtra	..	6·01
10. Orissa	1·03	3·20
11. Punjab	1·98	3·68
12. Rajasthan	3·15	8·73
13. Uttar Pradesh	7·58	27·07
14. West Bengal	2·10	7·57
15. Mysore	..	5·79

†Budget Estimate. *Relates to Bombay.

Note: This is a bad table for comparison over years because of reorganization of states. But the object is to illustrate how the land revenue collections vary from state to state.

[Source: *Report of the Finance Commission*, 1957 and Reserve Bank of India Bulletin, May, 1966.]

TABLE 7.9

Growth of Land Revenue

(Rs. Crores)

Year	Amount
1921–22	29·08
1936–37	25·96
1938–39	25·40
1944–45	30·21
1951–52	48·00
1966–67*	120·48

*Budget Estimate.

[Source: Vol. I, Report of the Taxation Enquiry Commission, 1953–54, and Reserve Bank of India Bulletin, May, 1966.]

It would be wrong to think however that land revenue has been altogether stagnant. The various ways in which the land revenue receipts have been augmented particularly during the Second and Third Plan are as follows: (1) a straightforward increase in land revenue; (2) levy of surcharge; (3) levy of cesses on land revenue by or for local bodies; (4) taxes on commercial crops; (5) enhancement of irrigation or water rates; (6) increases on account of the abolition of intermediaries and refixation of land revenue; and (7) assessment of Inams (grants of land recognized by government) in states like Madras. Different states have adopted different methods, in accordance with their own judgment. But rather than going in for a simple and straightforward increase, the states have preferred other methods which will preclude additional burden on small land owners. A surcharge, for example, can be levied on holdings bigger than a stated size. Cesses on land revenue have also been levied by various states for purposes of education and health. As a whole, land taxation has increased to some extent consequent on the introduction of Panchayati Raj. Many of the cesses and a portion of land revenue itself have been handed over in a number of states to the local bodies for specific or general purposes. Some states have also empowered local bodies for specific or general purposes. Some states have also empowered local bodies to levy a surcharge on other taxes on land.[1] Thus the full impact of taxation on land can be understood only if taxation by or for local bodies is also taken into account. The abolition of the intermediaries has also resulted in increase in land revenue in states like Rajasthan, Uttar Pradesh, Bihar and Madhya Pradesh

Ever since of the report of the Taxation Enquiry Commission it is generally agreed that the rural sector is comparatively undertaxed and that land revenue requires enhancement. Various proposals have been mooted for the rationalization and enhancement of land revenue.[2] In such a context the proposal to abolish land revenue put forth by the Madras government and keenly taken on by some other state governments,[3] has set things in an entirely different direction. It is no doubt true that land revenue procedures

[1]In Madras, since 1962–63, panchayat samithis (unions) have levied a surcharge which is, on an average, 25% of the land revenue.

[2]See, for example, *Basis of Taxation in the context of developing Indian Economy* (Rapporteur R. J. Chelliah), 1965, Popular Prakashan, Bombay.

[3]Put forward by the Congress Government before the Fourth General Elections and taken up by the D.M.K. which was voted to power.

have become quite cumbersome, the cost of collection comparatively high and the accounts complicated. But the main reason seems to be political. It is not our intention to discuss the question in all its details here. We have already seen from Table 8 the extent to which individual states are dependent on land revenue. But we may stress that if land revenue is to be abolished, the abolition should not extend to the cognate fields of water charges and betterment levies which will have to remain to meet the expenditure on the construction and maintenance of irrigation sources. The case is actually for an enhancement of these rates. Besides, the collection of land revenue and its enforcement through the Revenue Recovery Act has been a source of power to the Revenue Department in dealing with the public and this power has been used some times in beneficial ways to get the cooperation of the public to the schemes of government. The withdrawal of this power would mean a certain handicap to the administration in its effectiveness over the rural population.

It will not be easy for states to fulfil their targets of additional taxation for the Fourth Plan, if, besides, they have to bridge the gap created by a total or partial abolition of land revenue. As we have stated already, total abolition will not be necessary. The expedient course for the government would be to abolish land revenue (other than irrigation charges and cesses) as a state tax but allow local bodies to levy the tax according to local needs and circumstances. After all, land revenue is a very good local tax and should have been left to the exploitation of the local bodies by now. As we are going to argue elsewhere[1] there is a notorious imbalance between the functions and finances of local bodies and the delegation of land revenue at this stage would kill two birds with one stone. On the one hand, local finances can be strengthened and on the other the expenditure that would have to be incurred by way of grants in lieu of erstwhile devolutions of land revenue will be saved, thus providing an expedient solution to the state government. The opportunity can also be availed of for an appraisal of the consequential changes in the financial arrangements with the local bodies. The centre as well as the Planning Commission will no doubt take an adverse view of the states giving up land revenue altogether and probably the delegation of land revenue to the local bodies will be a compromise which may partially satisfy them. At

[1]Chapter XIII.

the local level too there will be a much greater willingness to pay taxes in order to secure particular local benefits.[1]

The centre can also avail of this opportunity to get the consent of all the Chief Ministers for making agricultural income taxable by the centre so that a consolidated income tax can be levied. It has long been a lacuna in Indian public finance that taxation of agricultural income has been left to the states while taxation of non-agricultural income vests with the centre. The Taxation Enquiry Commission as well as others have suggested a merger of these two income sectors. The present circumstances in which the states are prone to 'abolish' land revenue may be propitious for getting the consent of the Chief Ministers as a sort of compromise to transfer agricultural income tax to the centre for levy and collection and for eventual distribution to states on some simple and agreed formula.

It is necessary to have an idea of the taxation of agricultural income in India before we examine this issue further. Agricultural income-tax has been levied in most of the states in India but not with any particular success.[2] Where it yields comparatively substantial revenue the lion's share of it is derived from the plantations. The rates of taxation have been low and facilities for composition liberal. However, the implementation of land reform measures, particularly the imposition of land ceilings, has resulted in the sub-division of lands and affected the tax base of agricultural income tax very adversely. There is a very strong case for better exploitation of agricultural income tax from financial as well as equity considerations. The imposition of cesses and surcharges over land revenue has made the land tax quite regressive and a measure of agricultural income tax would restore the balance towards equity. On this ground alone an increase in agricultural income tax either by the state or by the local bodies in the form of a surcharge, would be very beneficial. But it is highly doubtful whether in the wake of the proposal to abolish land revenue states will consider seriously an upward revision of the agricultural income tax. In this context therefore there is much to be said for the central government taking the initiative and getting the concurrence of the state governments to merge

[1]Land revenue can be a good local tax but it will be regressive as between areas, since local bodies with lower land revenue will have to levy higher rates. That is why agricultural income tax needs to be used as a corrective.

[2]For 1966–67, the budget estimate for agricultural income tax for all the states was Rs.10.58 crores as against Rs.120.48 crores for land revenue. In 1963 as much as 55% of the total cropped area in the country was practically outside the purview of this tax. Cf. 'Agricultural Income Tax in India', Reserve Bank of India Bulletin, August, 1963, which gives details for the various states.

agricultural income in the total income for purposes of taxation.

This would of course hit those states which are deriving substantial revenue from plantations because the taxes from agricultural income would also become part of the divisible pool. Their interests would require to be protected by guaranteeing their current proceeds or by some other formula.

There has been a phenomenal increase in the number of wells in the country and there are special schemes for speeding up well irrigation. Probably a local tax on wells at moderate rates may give some revenues to the local bodies.[1]

The assessment of non-agricultural land assumed special importance with increasing urban development. The practice in this regard varies considerably from state to state. There are states which have made statutory provision for the assessment of non-agricultural land, other who levy it under executive orders and yet others who do not levy it at all. There is also great diversity in the basis and rates of assessment, the basis being annual value, gross rental value, market value or freehold value. The rates in many states are rather nominal. But a tax on a non-agricultural land and particularly, an urban land tax should be capable of benefiting from the rapidly rising land values in urban areas. The majority of the states have not however bestowed adequate attention to the taxation of unearned income derived from lands put to non-agricultural use. In the course of the Third Plan some states have had recourse to urban land tax and it is likely that others may follow suit. The Report of the Committee of Ministers on the augmentation of the resources of urban local bodies has suggested that the proceeds of such taxes should be given over to local bodies. We shall be discussing this aspect while considering it in the context of the additional taxation efforts of the states.

Another important source of revenue is 'stamps and registration'. Under registration, fees are levied for registering documents. The fees vary from state to state and are generally on an *ad valorem* basis. The fees on account of registration actually exceed the expenditure thereon substantially. To this extent, the fees have taken the character of a tax. Stamp duties in India are divided into judicial and non-judicial. The former are levied under the Court Fees Act 1870[2] and represent fees payable by persons having business in law

[1]The British system of land revenue settlement precluded taxation of improvements like wells but a moderate tax is not going to curb the cultivators' keenness to sink wells.

[2]This Act was recently struck down by the Madras High Court on the ground that the fees were out of proportion to the services rendered.

courts and public offices. Non-judicial stamp duties are regulated under the Indian Stamp Act, 1899 as amended from time to time by the Government of India and the state governments. The stamp duty on bills of exchange, cheques, letters of credit, promissory notes, receipts etc. is within the legislative competence of the centre but the proceeds are collected and appropriated by the states. Stamp duties on other items including transactions on immovable property are levied by states. Some stamp duties are fixed and others are on an *ad valorem* basis. Some states like Madras also levy a surcharge on the stamp duty on transfer of immovable property, to be collected by the state and handed over to local bodies.

The revenues from stamp duties and registration have shown a fair buoyancy for a number of reasons. The growing economic activity, the increasing level of prices and transactions in land consequent on land reforms have all contributed to this trend. States have not also been unwilling to raise the rates of stamp duties from time to time. In the First Plan a number of states revised court fees upwards and during the Second and Third Plans non-judicial stamp duties have been increased.

We may next take up taxes relating to motor vehicles. Motor vehicles are the objects of taxation at all levels, central, state and local; at the central level in the form of import duties and excise duties, at the state level in the form of sales taxes and motor vehicles taxes and at the local level by tolls and wheel taxes. We are concerned only with the state taxes relating to motor vehicles. Under the Constitution, while the taxation of motor vehicles is in the state list, the subject 'mechanically propelled vehicles including the principles on which taxes on such vehicles are to be levied' is included in the concurrent list. No legislation in regard to these principles has so far been enacted by the Union. At the state level we may consider three types of taxes relating to motor vehicles, namely, the motor vehicles tax as such, taxation on passenger and goods and the sales tax on motor spirits. The basis of motor vehicles tax as well as the rates vary from state to state. In some states goods vehicles are taxed on the basis of unladen weight, in some on laden weight and in a few others on authorized loads. Private motor cars are taxed either on unladen weight or according to seating capacity. Passenger buses are generally taxed on the basis of their seating capacity. During the First and Second Plans many states imposed taxation of passenger fares and goods freights, the tax being a percentage of the fare or the freight. Motor Spirits Tax has been in existence from 1937 and now most states levy this tax on the basis of a separate enactment. The rates of the tax vary from state to state. All these three types of taxation relating to motor vehicles have been resorted to increasingly by the states particularly in the Third Plan and the rates have become fairly high.[1] The taxation is said to be quite burdensome in a number of states and it has been calculated that more than 50% of the cost of the operation of the buses is accounted for by some taxes or other.[2] This source of revenue has been rather tempting to the states and it has proved to be elastic too. The revenue is also easy to collect. But the future elasticity of the taxes relating to motor vehicles is limited by the overall target production of automobiles in the country. From the point of view of the finances of the state governments there is a very good case indeed for a larger production of automobiles! The benefits of automobile production are not however evenly spread because states like Maharashtra, West Bengal and Madras use a larger proportion of vehicles than others.

There have been difficulties in inter-state transport because of disparities in taxation and these difficulties are usually resolved by interstate agreements. Sooner or later, central intervention or constitutional amendment is bound to become necessary.[3]

Mention may also be made of the tax on entertainments and the show tax which are levied by a number of states. The entertainment tax is generally levied on the prices charged for admission to any place of entertainment. The bulk of the revenue from this tax is provided by the cinema. The show tax is a tax for each show conducted by the exhibitor and it falls on him, while the burden of the

[1]In 1963–64 the variations were as follows: (1) Public carriers, Rs.300 in Jammu and Kashmir to Rs.3,200 in Madras. (2) 40 seater buses, Rs.320 in Jammu and Kashmir to Rs.10,800 in Andhra. (3) Tax on passenger fares, 10% in a number of states to 17·5% in Gujarat. (4) Tax on freights, 3% in Gujarat and Maharashtra to 12·5% in Bihar. See Report of the Committee on Transport Policy and Co-ordination, Planning Commission, 1966.

[2]As regards trucks, 'the Indian Road Transport Development Association has estimated that each truck on the road pays on average every year taxes totalling about Rs.11,000, including taxes and duties on diesel oil which alone amount to Rs.7,500. This is not all. In return for the crushing taxes that the operator pays, he is rewarded by a grossly inadequate mileage of badly constructed and poorly maintained roads, by delays and corruption inherent in a multitude of octroi posts and weigh-bridges, and by a system of permits, regulations and restrictions which make the crossing of the borders of a district or a state more difficult and time consuming than crossing the national frontiers in other parts of the world', *Capital*, 4th August, 1966, page 253.

[3]See Report of the Committee on Transport Policy, already cited.

entertainment tax is borne by the cinema-goers. The entertainment tax is collected either in cash or by special adhesive stamp. Problems of evasion arise in comparatively large magnitudes in respect of the entertainment tax. In a number of states the entertainment tax collected is passed on to local bodies. This would appear to be the correct position since the tax is essentially a local tax. In Madras state, apart from the basic entertainment tax levied by the state, the local bodies themselves have been empowered to levy a surcharge up to 100% of the basic tax on entertainment and 300% on the show tax. A number of local bodies have availed themselves of this opportunity and this has provided them with a further source of revenue which is quite considerable in urban areas.

State excises are confined to excise duty on liquor (except medicinal and toilet preparations) and narcotics. As is well known, the introduction of prohibition considerably whittled down the importance of the tax which was one of the pillars of the state finances previously. There has however been some rethinking on prohibition policy though it is not likely that many states will scrap prohibition.[1] However states which have not introduced prohibition have in recent years increased the excise duties.

A tax which can produce large revenues and impart a substantial measure of elasticity to the financial structure of the states is the tax on consumption of electricity. 'Electricity duties' have been in vogue in many states for a long time but they are duties on the electricity sold by private licensees to consumers. This will not include the electricity sold by Electricity Boards. A more comprehensive and productive tax is a tax on the consumption of electricity by consumers whether it is sold by private licensees or by Boards. In Madras this tax has proved to be a very good source of revenue. Any effects it may have on industry can be regulated by classification of the industries and by levying separate charges for high tension and low tension and so on. A tax of this kind needs to be urgently introduced in all states. India is still in the early stages of power generation and hence the potentialities of this tax are immense. We shall be discussing this again in the next chapter in the context of additional taxation.

There is a related aspect here, namely, that to the extent that there is a tax on consumption of electricity there may be a limit to the revision of electricity tariff by the electricity boards for their

[1]Kerala has announced withdrawal of prohibition. The increase in revenues is expected to be about Rs.7 crores (exclusive of savings in cost of enforcement). Mysore, Madhya Pradesh and Haryana have also announced such withdrawal.

own purpose. Studies have revealed that electricity boards themselves have to go a long way to financial stability.[1] There is thus the dilemma whether the electricity boards should be allowed to raise the tariff and thus have a sound financial base or whether the state should appropriate such increases for its own purposes. Obviously in most cases the dilemma will have to be resolved by partial increases in both or depending on which needs the resources badly, the electricity board or the state. After all, in a country like India, electricity cannot be said to be very expensive and it can be made a little expensive when it is really giving substantial benefits to the people. The difficulty will arise only in respect of electricity-intensive industries for which suitable concessions can be given. The treatment of agriculture may also involve some difficulties because many states may be averse to increasing the rates for supply of electricity to agriculturists.[2] But this will have to be done sooner or later. If necessary, the proceeds from the increase in agricultural tariff can be credited to some kind of revolving fund from which money may be spent for extension of power supply for agricultural purposes.

Before we take up sales tax, we may also mention, without any discussion, a number of miscellaneous taxes and duties which are levied by some states or other, like: forward contracts tax (Punjab), purchase tax on sugar (Madhya Pradesh, Madras and Uttar Pradesh), sugar cane cess (Andhra, Bihar, Mysore, Madras, Uttar Pradesh), tobacco duties (Kerala, Madhya Pradesh, Mysore), taxes on prize competitions and betting (Andhra, Gujarat, Madras, Maharashtra, Mysore, Uttar Pradesh, West Bengal), advertisement tax (Gujarat and Maharashtra), taxes on buildings (Kerala), tax on raw jute (West Bengal), tax on vacant plots in urban areas (Rajasthan) and tax on entry of goods (West Bengal).

We turn finally to the place of sales tax in state finances. The subject of sales tax in all its ramifications, like its formal and effective incidence, its administration, its complexity and disparities among states, deserves a book by itself. Data on its rates and structure in the various states are not easily available for a comparative study. As a matter of fact, even in the Fourth Finance Commission which was requested to go into certain aspects of the sales tax could not get the simple information relating to the rates prevalent for certain commodities in some states like Assam, Jammu and

[1]See also Chapter XII.

[2]Many states are even now charging less than the optimum tariff for agriculture.

Kashmir, Rajasthan and West Bengal. The information was said to be 'not readily available'.

Thus a description of the system of sales tax in the states in India is beset with difficulties. Following the Taxation Enquiry Commission we may classify some of the common methods of sales taxation in India.[1] First we must distinguish between a sales tax and a purchase tax. The sales tax is a tax which with reference to the transaction of sale-cum-purchase is levied on the seller whereas the purchase tax is a tax which with reference to the same transaction is levied on the buyer. Another distinction is between a selective sales tax and a general sales tax, the former being on selected goods and the latter on the large majority of goods sold by the dealers. It is the general sales tax that is important and it has two main varieties, multi-point and single-point. There is also a double point tax (as in Bombay) but it is regarded as a variation of the single point system.

An important distinction in conception and design between the multi-point system and the single-point system is that the latter seeks to ensure as far as possible that not more than a specific amount of tax gets added to the price at which the article is sold to the consumer: that is, that the total tax charged in the passage of the article from the first dealer to the last in a chain of dealers is not left to be decided by the indeterminate number of links which may exist in the chain but is collected at only one of the number of sales which may be involved. The single-point tax is usually levied at a relatively high rate and has a large number of exemptions. But the multi-point system is not incidence-controlled in structure or purpose. The rate is relatively low and the exemptions few.

The actual history of sales tax in India has witnessed two contrary initial trends. States like Madras began with a low multi-point tax and states like West Bengal began with a comparatively high single-point tax. Developments in either case have however been such that a combination of single-point and multi-point system has become the feature of most states. The multi-point system may be said to be more applicable to a state with considerable rural area and population. It is occasioned by a decision of government (1) to make the tax applicable to the majority of articles of consumption (including those which may be relatively essential); (2) to levy it on a large number of dealers (big, medium and small) and at the same time; (3) to levy so low a rate that it is not a matter of serious

[1]The description of the sales tax system in India follows that of the Taxation Enquiry Commission, 1953–54, Vol. III.

consequence that tax is paid several times over on the same articles or that it is also levied on essential articles.

On the other hand the single-point tax may be said to follow from a decision of the following nature on the part of government: (1) to make the tax applicable to a large variety but not the majority of articles; (2) to levy it on a limited number of dealers on the underlying assumption that the articles in question are such as will by and large pass at one stage or other through the limited but defined sector of the tax-paying dealer and (3) to pitch the tax at a relatively high rate, ensure that it does not have to be paid twice over and, further, since some of the articles sold by these dealers cannot well be taxed at that rate because of their importance to the less well-to-do consumers, provide for an adequately wide range of exemptions.

Though there are criticisms of single as well as multi-point systems, it has by now been recognized that from the financial, economic and administrative points of view, only a composite tax system, where both the taxes have a part to play, will be the practical solution. Sales tax administrations have inevitably acquired a measure of complexity. Complications can also arise in regard to foreign trade and inter-state trade. The Constitution prohibits sales taxation of goods involved in foreign trade. It provides that no state shall impose 'a tax on the sale or purchase of goods where such sale or purchase takes place (a) outside the state; or (b) in the course of import of the goods into, or export of the goods out of the territory of India'. In spite of the provision, a number of difficulties were experienced in inter-state trade. The Taxation Enquiry Commission also spotlighted the difficulties in this regard and considered that inter-state sales should be the concern of the Union and the sphere of power and responsibility of the state may be said to end and that of the Union begin when the sales tax on the one state impinges administratively on the dealers and physically on the consumers of the other states. The Commission also suggested the amendment of the Constitution. In 1956, the Constitution was amended and a new item was introduced in the Union List: 'Taxation on the sale or purchase of goods other than newspapers where such sales or purchase takes place in the course of inter-state trade or commerce'. Parliament was empowered to formulate by law principles determining when a sale or purchase of goods takes place in the case of inter-state trade or import or export. It was also entrenched with the law that a state law in so far as it imposes a tax on the sale or purchase of goods declared by Parliament by law to be of a special importance in inter-state trade or commerce shall be subject to such restrictions and conditions as Parliament may specify. Parliament also enacted the Central Sales Tax Act in 1957 and the Central Sales Tax was imposed from the first July 1957. This Act enumerated 'declared goods' which are of special importance to inter-state trade or commerce. It laid down the principles for determining when a sale or purchase takes place in the course of inter-state trade or import or export. It provided for the levy, collection and distribution of taxes on the sale of goods in the case of inter-state trade. The tax was originally levied at one per cent and is now three per cent. The imposition of the Central Sales Tax has further complicated the sales tax system but it has at the same time introduced a measure of uniformity and obviated the conflicts and confusion that are inevitable in inter-state sales.

There is no doubt about the crucial role played by the sales tax in state finances, providing for a measure of elasticity and buoyancy to state revenue. Table 7.10 will show the growth of sales tax as compared to national income and Union excise.

TABLE 7.10

Growth of National Income, Union Excise and Sales Tax

(Rs. Crores)

Year	Revenue from Union Excise	Revenue from Sales Tax	National income at current prices	Excise Revenue as percentage of national income	Sales Tax Revenue as percentage of national income
1950–51	67·5	60·0	9,530	0·7	0·6
1951–52	85·8	59·0	9,970	0·9	0·6
1952–53	83·0	55·6	9,820	0·8	0·6
1953–54	95·0	62·6	10,480	0·9	0·6
1954–55	108·2	71·6	9,610	1·1	0·7
1955–56	145·3	78·0	9,980	1·5	0·7
1956–57	190·4	84·6	11,310	1·7	0·7
1957–58	273·6	117·3	11,390	2·4	1·0
1958–59	312·9	123·9	12,600	2·5	1·0
1959–60	360·7	136·8	12,950	2·8	1·1
1960–61	416·4	158·8	14,140	2·7	1·1
1961–62	489·3	181·4	14,800	3·3	1·2
1962–63	598·8	208·9	15,400	3·8	1·4
1963–64	729·6	268·3	17,200	4·2	1·6
1964–65 (RE)	773·1	297·5	..	..	..
1965–66 (BE)	819·2	315·9	..	..	..

[Source: p. 232, *Report of the Finance Commission*, 1965.]

The growth of sales tax has been higher than the growth of national income but Union excise has grown even faster. Though, even as it is, sales tax has great potential for elasticity and buoyancy, the revenue would be more if the possibilities of evasion are checked. There appears to be no general estimate on the loss of revenue to states because of evasion but considering the fact that states have from time to time taken credit for substantial amounts as receipts by way of 'tightening up of sales tax system', evasion does seem to absorb large revenues. Estimates of loss due to evasion however vary widely. In view of the limited possibilities for increase in revenues for the states, it is quite necessary that all methods of evasion are checked. The states could do well to compare notes on methods of checking evasion.

TABLE 7.11

Sales Tax as a proportion of Tax Revenue

(Rs. Crores)

States	1965–66 (Budget Estimates)		
	Total revenue from Sales taxes*	Total tax revenues of the States (excluding share of the central taxes)	Sales tax revenue as percentage of total tax revenue
1. Andhra Pradesh	22·0	65·4	34
2. Assam	7·4	25·0	30
3. Bihar	18·5	50·1	37
4. Gujarat	22·6	44·4	51
5. Jammu and Kashmir	0·9	3·4	25
6. Kerala	17·4	37·0	47
7. Madhya Pradesh	18·9	51·5	37
8. Madras	34·8	72·2	48
9. Maharashtra	60·9	110·5	55
10. Mysore	17·2	43·0	40
11. Orissa	8·3	18·8	44
12. Punjab	17·0	50·0	34
13. Rajasthan	11·7	23·7	35
14. Uttar Pradesh	22·6	82·8	27
15. West Bengal	35·7	77·2	46
All States	315·9	764·7	41

*Includes General Sales Tax, Central Sales Tax and Tax on Motor Spirit.
[Source: p. 233, *Report of the Finance Commission*, 1965.]

The importance of the role of the sales tax in the finances of the various states can be seen from Table 7.11. It will be seen that sales tax constitutes on an average as much as 41% of the total tax revenues, the percentage in individual states varying from 25 in Jammu and Kashmir to 55 in Maharashtra. Generally speaking the somewhat better industrialized states like Madras, Maharashtra and West Bengal seem to have greater reliance on sales tax.

As has been pointed out by the Sales Tax Enquiry Committee of Bombay, 'Sales Tax owes its present position—not merely to the size of its yield, but to the expansiveness which its inherent elasticity and wide coverage imparts to the fiscal structure of the state. . . . By comparatively simple process of adjustment of the rates and incidence of the sales tax . . . or by modification of the structure of the tax . . . the state can, with singular efficiency, regulate its revenues within fairly wide limits'.[1]

As the Taxation Enquiry Commission observed, the merit of the sales tax as a major source of revenue lies in the fact that it is dispersed over a large number of goods and of dealers, and this makes possible the realization of substantial revenue from a comparatively low rate of tax, whereas, by definition, customs and excise apply only to a strictly limited portion of the industrial output that is sold within the country. Even if the sales tax was levied not by the states as at present, but by the central government on an all-India basis, it would not, in the circumstances mentioned, be a superfluous item in the tax system which could be conveniently and effectively replaced by a combination of excise, customs and octroi.[2]

The future lines of action in regard to sales tax will have to be in the form of rationalization and reduction of evasion. States have not been adequately conscious of the need for systematic study of the role of sales tax in state finances and the manner in which the procedures and administration could be improved. Useful data on sales tax are hard to come by and an inter-state study of sales tax is a major desideratum in state finances.

The place of sales tax in the fiscal system of the Union and the states is an area of debate which never remains quiet. Apart from inter-state sales, problems also arise relating to the roles of sales tax and central excise and their impact on trade.

The Fourth Finance Commission was asked specifically to

[1]Report of the Bombay Sales Tax Enquiry Committee, 1957–58.
[2]Page 44, Report of the Taxation Enquiry Commission, 1953–54, Vol. III.

consider, the effect of the combined incidence of a state's sales tax and Union duties of excise on the production, consumption or export of commodities or products, the duties on which are shareable with the states, and the adjustments, if any, to be made in the states' share of Union excise duties if the sales tax rates levied by the states exceed a specified ceiling.

In the absence of adequate data, the Finance Commission was unable to measure the combined incidence of excise duties and sales tax. Being so unable, the Commission considered it inadvisable to proceed to discuss the question of fixation of any ceiling. Even if it had proceeded to the question of fixation of ceiling, it would have been sailing on difficult waters, since fixation of ceiling for individual commodities is beset with problems of its own. The real point is the sales tax is the backbone of the financial structure of the state governments and their only ray of hope in difficult situations. Any attempt to place restrictions on the freedom of the states in the field of sales tax would affect their capacity to raise resources. It is also almost the only major instrument left with them for shaping the economic and industrial policy in their area. Now that major question of taxation of inter-state sales has been resolved more or less satisfactorily, the states can be left to exploit sales tax to their own advantage. The Taxation Enquiry Commission categorically said that sales tax was to remain with the states and cannot be abolished by an extension of customs, excise, or octroi and that centralization of sales tax must be ruled out; the states cannot do without sales tax.

This is of course not to rule out co-ordination of sales tax policies in the interests of the economy as well as the taxpayer. Occasions will arise in future too when co-ordination will be necessary. The decision to give up the levy of sales tax on certain commodities and their substitution by additional excises by the centre (for eventual distribution to states) is a very good example of such co-ordination. Another example is the levy of a uniform rate of 10% of single point sales tax on certain luxury items made by all states on the suggestion of the Union Finance Minister in 1963. The substitution of sales tax by additional duties of excise has not however gone without controversy, as we have seen already.

The heart of the matter is really that there is no proper institutional arrangement for a study and co-ordination of sales tax policies among states and between the states and the centre. We may therefore reiterate the need for inter-state research on this subject preferably by a tax research council, which can be either an independent body or part of an inter-state institution with a larger scope.

While on the point of state taxes we should also stress the need for concerted action on the part of the states to collect the taxes without arrears. As at the end of 1963–64, the arrears of revenue were Rs. 28 crores in Uttar Pradesh, Rs. 16 crores in Andhra Pradesh and West Bengal, Rs. 14 crores in Maharashtra, Rs. 10 crores in Bihar and Kerala, Rs. 9 crores in Rajasthan, Rs. 8 crores in Mysore, Rs. 7 crores in Madras and Jammu and Kashmir, Rs. 6 crores in Assam, Rs. 5 crores in Gujarat, Rs. 4 crores in Orissa and Rs. 3 crores in Punjab.[1]

Before we conclude this chapter we may make a brief mention of non-tax revenues (other than grants-in-aid). Non-tax revenues have also grown over the years though, as we have seen, their proportion to the total revenues of the states has not changed appreciably. This revenue consists of administrative receipts, net contribution of public enterprises like forests, irrigation, electricity schemes, road and water transport, industries and others, other revenues like interest payments from borrowers, stationery and printing receipts and miscellaneous receipts. It is not necessary to discuss these components in detail because the conditions vary from state to state and we are in any case dealing with important aspects like irrigation and electricity schemes elsewhere. We could however stress at this juncture that in no state (except probably Madhya Pradesh) has sufficient attention been given to the components of non-tax revenue and the means of increasing them. Apart from the irrigation and electricity schemes, there are possibilities of increased revenue in respect of other schemes and also by revision of licence fees etc. For this purpose the items of non-tax revenue will have to be scrutinized one by one. The potentialities under each item will vary from state to state. We shall be mentioning in the next chapter, some of the measures of non-tax revenue that states have adopted as part of their effort in raising resources and how this effort has also varied from state to state.

It would hardly have escaped notice that our discussion of state taxes has not brought in the theoretical considerations of Public Finance. Questions relating to incidence and effects of state taxes have been left untouched. The reason is that a lengthy discussion of these aspects (and they cannot be dealt with briefly) will be outside the scheme and approach of the book. Besides, there is likely to be not much point in considering the incidence and effects of state

[1]See Report of the Finance Commission, 1965, pp. 218–9.

taxes alone in a federal economy. In fact there is need for a thorough-going study of the matter but the Taxation Enquiry Commission as well as the Finance Commission have been unable to go in for these studies for lack of data. Taxes, however, are mostly levied on *ad hoc* and empirical considerations. It is not as if in financial administration a careful study of the incidence and effects of taxation is available before the administrator and the politician at the time of formulation of proposals for additional taxation. Taxation efforts are ultimately governed by the confidence of the political executive to carry them through, and by administrative considerations relating to the cost of administration and the wieldy or unwieldy nature of the tax administration and coverage. How the states have set about the task of additional taxation we shall see in the next chapter.

CHAPTER VIII

TAX EFFORTS OF STATES

We shall devote ourselves in this chapter to an examination of the efforts of additional taxation made by various states in fifteen years of planning. A separate consideration of this is necessary for more than one reason. That additional resources have to be raised by states is not in doubt. Complaints are heard from time to time that while the centre has been fulfilling its quota of additional taxation, the states, as a class, have not pulled their weight. Particular states have been remarkably 'restrained' in matters of additional taxation while there are others which have gone about the matter of additional taxation in a businesslike way. The discussion can also help us to understand whether there is or there is not, scope for increasing state taxes and if so, the nature of increase that the states have found feasible.

We may mention here that it is very difficult to calculate exactly the additional revenue due to additional taxation. Such increases as are due to enhancement of existing levies and not due to imposition of new levies or an extension to items which can be separately identified are difficult to arrive at. Besides, the additional taxation efforts we are going to consider are the proposals outlined in the Budget Speeches as reported in the Bulletin of the Reserve Bank of India every year. Actual implementation might have been somewhat different.[1] It is also possible that some minor measures without an identifiable quantitative significance may have been omitted.

There is no doubt that the states did very badly by way of additional taxation during the First Plan. As against a target of Rs. 230·30 crores for the Plan period their achievement was only 80·4 crores, in other words a percentage of 35.[2] Table 8.1 will show the quantum of additional taxation proposals made by each state during each year of the Plan.

It will be seen that substantial taxation proposals were undertaken only during 1952–53 and 1953–54 and no state appears to

[1]For example, certain proposals of Bihar in the Second Plan.
[2]P. 35, *Report of the Finance Commission*, 1957.

have made very concerted attempts in carrying through a programme of additional taxation. West Bengal had practically no additional taxation at all even though it had a target of nearly Rs. 37 crores to fulfil. Madhya Pradesh also had no taxation worth the name and so also Orissa.

TABLE 8.1

Taxation Proposals in First Plan

(Rs. Lakhs)

States	1951–52	1952–53	1953–54	1954–55	1955–56
Andhra Pradesh	—	—	—	100	—
Assam	—	—	—	NS	—
Bihar	—	—	60	—	—
Bombay	—	255	—	—	—
Madras	—	270	50	40	170
Madhya Pradesh	—	—	—	—	—
Orissa	—	—	—	NS	—
Punjab	—	18	—	61	—
Uttar Pradesh	—	800	364	—	—
West Bengal	—	—	—	—	—
Hyderabad	—	224	12	10	40
Madhya Bharat	—	15	21	—	—
Mysore	25	—	52	100	20
Pepsu	—	—	—	15	13
Rajasthan	—	—	15	—	—
Saurashtra	3	—	—	—	—
Travancore-Cochin	2	—	6	—	26
TOTAL	30	1,582	580	326	269

NS: Details not stated.

Note: This has been tabulated from the taxation proposals mentioned in the Bulletins. It is possible that some proposals have been omitted. In 1953–54 there were tax concessions to the extent of Rs. 7 lakhs in Punjab.

[Source: Reserve Bank of India Bulletins.]

Of the additional Rs. 80·4 crores raised by states during the five years about 50% was accounted for by sales tax and taxation of motor spirit and tobacco and 20% by motor vehicles, passenger and carriage taxes. Taxation on land contributed very little and except in Uttar Pradesh irrigation rates also did not contribute any sizable amount. In sales tax, the major event was the change over by Bombay from single point to multi point yielding an annual revenue of 2·5 crores. Some amplification of sales tax system was also made, among others, in Madras and Andhra states. This period also witnessed the levy or enhancement of a tax on the sale of motor spirit. Another feature was the introduction of a tax on passenger fares and freights in regard to motor vehicles, particularly by Madras and Punjab, in 1952–53. This was followed by other states during the First and Second Plans. Yet another feature was the levy of a surcharge on agricultural land and also more intensive taxation of plantations. Betterment levy was imposed in a number of states though it is not clear whether it was actually collected, because the course of betterment levies has never run smooth. As a matter of fact it may be mentioned here that concerted action by the Planning Commission during the First Plan and the early years of Second Plan to increase land revenue and enforce betterment levies might have resulted in a more orderly and fruitful implementation leading to greater revenues. Agricultural income tax on plantations also came to be levied in Madras and Mysore and Orissa. Yet another feature was the revision of court fee schedules in many states. Not much revenue was derived by enhancement of stamp duties.

Taking the states as a whole there was clearly a lack of systematic effort at raising revenues during the First Plan. The upshot was the poor performance by a number of states, as will be seen from Table 8.2

The target for additional taxation by the states for the Second Plan was fixed as Rs. 225 crores and it was expected to be realized roughly in the manner indicated in Table 8.3.

The achievements of individual states are not available. The taxation requirements seem to have been increased in the course of the Plan. The Third Plan simply states that additional taxation efforts by the centre and the states exceeded the estimates. Perhaps it ought to be taken that the states as a class had not fallen short of expectation. In the first three years they had levied taxation which would give a yield of Rs. 123 crores over five years as against the targets of Rs. 225 crores.

Table 8.4 will show the efforts undertaken by individual states. Most states undertook some taxation effort in 1956–57. Bihar proposed certain increases expected to yield Rs. 2 crores per annum. The taxation effort of Orissa and West Bengal for the period as a whole was poor. Andhra Pradesh, Assam and Mysore made some modest tax proposals for only one year. Hence even if the states as a whole had done well it may be stated without fear of contradiction that there were individual states whose performance left much to be desired.

TABLE 8.2

Taxation Achievements in First Plan

(Rs. Crores)

State	Five Year Target	Achievement 1951–56	Achievement as a percentage of target
Assam	3·5	3·3	94·29
Bihar	7·3	3·0	41·10
Bombay	23·5	24·0	102·13
Madhya Pradesh	10·6	2·3	21·70
Madras (including Andhra)	39·3	8·0	20·36
Orissa	9·4	2·0	21·28
Punjab	5·0	4·5	90·00
Uttar Pradesh	50·2	11·0	21·91
West Bengal	36·9	4·5	12·20
Hyderabad	7·4	1·0	13·51
Madhya Bharat	4·9	2·7	55·10
Mysore	9·2	3·0	32·61
Pepsu	4·1	0·4	9·76
Rajasthan	3·3	2·6	78·79
Saurashtra	4·7	2·1	44·68
Travancore-Cochin	11·0	6·0	54·55
TOTAL	230·3	80·4	34·97

[Source: *Report of the Finance Commission*, 1957, p. 28.]

TABLE 8.3

Planned Taxation Proposals in Second Plan

(Rs. Crores)

Land Revenue	37·0
Agricultural Income Tax	12·0
Betterment Levy	16·0
Irrigation Rates	11·0
Sales Tax	112·0
Electricity Duty	6·0
Motor Vehicles Tax, Stamp Duties and Court Fees, etc.	14·0
Others (mainly local property taxes)	17·0
TOTAL	225·0

[Source: Second Five Year Plan, p. 89.]

TABLE 8.4

Actual Taxation Proposals in Second Plan

(Rs. Lakhs)

	1956–57	1957–58	1958–59	1959–60	1960–61
Andhra Pradesh	21	—	—	—	—
Assam	21	—	—	—	—
Bihar	200	25*	30*	—	—
Bombay	—	—	300	125	—
Madras	145	150	170	—	—
Madhya Pradesh	164	—	—	50	—
Orissa	—	—	—	10	—
Punjab	73	98	244	72	—
Uttar Pradesh	550	133	—	70	—
West Bengal	—	—	—	—	—
Kerala	—	233	67	—	—
Hyderabad	70	—	—	6	—
Madhya Bharat	55	—	—	—	—
Mysore	—	—	—	50	—
Pepsu	24	—	—	—	—
Rajasthan	—	105	80	68	65
Saurashtra	8	—	—	—	—
Travancore & Cochin	—	—	—	—	—
Jammu & Kashmir	—	—	—	—	—
TOTAL	1,331	744	891	441	65

*Bihar repeated certain proposals of 1956–57 in 1957–58 and 1958–59.

Note: This has been tabulated with reference to the proposals mentioned in the Bulletins. It is possible that some may have been omitted. In 1959-60 there were tax concessions for Rs. 50 lakhs in Uttar Pradesh.

[Source: Reserve Bank of India Bulletins.]

The major sources of increase during this plan too were general sales tax and taxes on motor spirit and motor vehicles. Bihar introduced a general sales tax besides its single point sales tax. Bombay levied a tax on passenger fares and freights. There were increases in electricity duty also.

We now come to the performance of the states in the Third Plan.[1] A target of Rs. 610 crores was fixed for the states as a whole and they have achieved it. Table 8.5 will show the additional taxation proposals announced by each state during each year of the Plan.

[1]What follows is based on the author's article 'States' Tax Effort in the Third Plan' in *Economic Weekly*, (Bombay), 20th March, 1965.

TABLE 8.5

Taxation Proposals in Third Plan

(Rs. Crores)

State	1961–62	1962–63	1963–64	1964–65	1965–66	5 year target
Andhra Pradesh	—	5·00	6·00	—	—	53·00
Assam	0·15	1·58	1·60	0·86	—	16·00
Bihar	—	*	—	—	—	50·00
Gujarat	0·80	2·23	2·90	0·08	0·25	29·00
Jammu and Kashmir	—	1·10	1·00	0·75	—	8·00
Kerala	2·13	—	5·50	—	—	23·00
Madhya Pradesh	3·00	5·00	1·90	1·45	2·11	48·00
Madras	—	7·14	—	1·00	—	45·00
Maharashtra	1·00	5·71	0·90	—	—	52·00
Mysore	1·70	4·00	—	—	—	42·00
Orissa	0·69	1·57	0·70	—	—	23·00
Punjab	1·31	3·80	5·30	—	—	40·00
Rajasthan	1·35	0·15	4·50	0·58	1·73	32·00
Uttar Pradesh	5·21	†	0·30	—	—	109·00
West Bengal	—	—	6·50	1·90	—	40·00
TOTAL	17·33	38·41	37·00	6·62	4·14	610·00

*Taxation to the extent of Rs. 1 crore is reported to have been made in 1962–63 but no details are available.

†No details are available, though there was some taxation. Presumably the sum of Rs. 1·13 crores needed to tally the total was shared by Uttar Pradesh and Bihar.

[Source: Reserve Bank of India Bulletins.]

It will be seen that the performance of Bihar has been rather nominal. Speaking broadly, states like Madhya Pradesh, Punjab, Rajasthan, Maharashtra and Madras seem to have been rather 'progressive' in the introduction of tax measures without any hesitation whereas states like Bihar, West Bengal and Uttar Pradesh seem to have been rather reluctant to impose measures of additional taxation. In 1961–62 there was not much of additional taxation, obviously in view of the impending elections. There was a substantial doze of additional taxation by many states, in 1962–63 and the trend was more or less kept up till the end of the Plan.

The directions in which additional taxation was undertaken would be seen from Table 8.6. General sales tax and taxes on motor vehicles

TABLE 8.6

Tax Proposals in Third Plan—Tax Wise

(Third plan, Rs. Crores)

Tax	1961–62	1962–63	1963–64	1964–65	1965–66
Land Revenue and Water Charges	4·60	7·38	4·00	1·00	—
Agricultural Income Tax	—	*	*	0·20	—
Tax on Non-agricultural Land	0·45	1·32	1·40	0·20	—
Motor Vehicles Tax (including taxes on fares and freights)	3·08	9·31	6·20	0·03	0·87
Electricity Duty	0·50	3·29	2·80	0·08	0·25
General Sales Tax	4·49	4·60	11·00	2·84	0·41
Tax on Motor Spirits	1·65	0·89	—	—	0·27
Stamp Duty	0·18	1·85	2·20	0·55	0·12
Entertainment Tax	0·31	0·99	0·50	0·20	0·25
State Excise	1·30	0·15	3·00	0·07	0·55
Others†	0·30	8·63	3·30	0·95	1·38
Non-tax measures	0·60	—	2·60	0·50	0·04
TOTAL	17·73	38·41‡	37·00	6·62	4·14

*There were changes in agricultural income-tax in Assam, Mysore and Jammu and Kashmir but no break-up is available.

†Including those for which break-up is not available.

‡Subsequent modifications resulted in a loss of Rs. 1·63 crores.

[Source: Reserve Bank of India Bulletins.]

were again the spearheads of the tax efforts of the states. There was some increase in land revenue though not probably to the extent envisaged. Electricity duty was emerging as an important source of revenue.

In respect of land revenue the increase in resources took the form of straightforward increase in many states like Assam, Andhra Pradesh, Kerala, Madras and Mysore. There was some emphasis on irrigation rates and betterment levies in Uttar Pradesh, Orissa and Madhya Pradesh. A comparatively new feature was the introduction in Maharashtra of a per acre levy on cash crops in irrigated lands. Andhra Pradesh also suggested an increase in the levy on sugar cane lands. Gujarat proposed a levy on cash crops but withdrew it and substituted it with a straightforward increase in land revenue. Punjab also levied a tax on certain cash crops on a per acre basis with different rates for canal-irrigated and other areas. Another feature was the introduction of a tax on land holdings (related to their capital or annual value) of over 30 acres in Uttar Pradesh and over 45 acres in Rajasthan. In West Bengal an increase in land revenue by Rs. 1 crore has been ascribed to land reform measures. There has been no particular shift towards the use of agricultural income tax as a measure of taxing the rural areas. The diversity of the measures by which the taxation of land is sought to be increased is at once a measure of the possibilities and the reluctance of the state governments to go in boldly for greater taxation of land.

Stamps and Registration have shown a measure of buoyancy since the Second Plan, probably due mainly to land reform measures which resulted in legitimate and not so legitimate transfers of land. All states except Andhra Pradesh, Bihar, Jammu and Kashmir, Madhya Pradesh and Uttar Pradesh effected straightforward or differential increases in the rates of stamp duties. It is well known that in this sphere there is a certain amount of leakage of revenue due to undervaluation and it is therefore significant that Rajasthan has proposed to take steps to check evasion. Registration fees were stepped up in West Bengal, Orissa, Madras, Madhya Pradesh and Rajasthan.

Another feature of land taxation that ought to become increasingly important is taxation of urban land. Increasing urbanization lends scope for this taxation which however seems to be lagging behind its potentialities. In the Third Plan period Andhra, Assam, Madras, Mysore, Orissa and Uttar Pradesh have introduced this taxation.[1] Jammu and Kashmir has levied a tax on property of commercial establishments and Rajasthan has imposed a tax on vacant lands and a development fee. Gujarat, besides stepping up non-agricultural assessment and having 50 per cent higher rates for vacant plots, seems to have taken a step ahead by withdrawing exemption from assessment for lands in village sites adjacent to towns with a population of over 5,000. Bihar, West Bengal and Punjab which all contain increasing areas of urbanization do not appear to have taken any steps towards taxation of urban land.

The figures relating to this form of taxation are not encouraging and this may be partly due to the manner of exhibition of these

[1]Uttar Pradesh abolished it in 1967 as a fulfilment of an election promise.

figures in the accounts and partly due to the time taken to put the scheme on the field and begin collection. There is no doubt however that this can be and ought to be an increasingly important form of taxation. Here the financial inter-relationships between the state governments and the local bodies come into play. Taxation of urban property, whether regularly or at the time of transfer, comes into competition with property taxation by municipal authorities and the Taxation Enquiry Commission had recommended that the urban immovable property tax should be progressively reduced by the states levying the tax as and when and to the extent the municipality concerned raises the rate of its tax. Recognizing this, Gujarat decided to discontinue the tax in Ahmedabad, the corporation having agreed to raise the property tax with effect from the same date. Likewise, Maharashtra proposed to transfer the proceeds of this tax to the local bodies and did not take credit in the budget for receipts under this tax.

There is obviously more than one solution to this problem and the solution actually chosen in each state depends on local conditions. The possible solutions are: (1) The entire field of taxation of urban property can be left to the local bodies; (2) the state government can levy uniform rates, collect them and hand them over to the local bodies; (3) the state government may levy, collect and appropriate the tax and give an equivalent amount of grants to local bodies, introducing a measure of equalization in the process; and (4) the state government may levy such taxes and not accept any correlation at all to municipal taxable capacity or grant-in-aid. Suffice it to say that this question brings into play the entire gamut of state-local relations and the need to achieve a purposeful parity between functions and finances in local bodies.

We may now turn to motor vehicles taxation, including taxes on fares and freights as well as motor spirits. This is becoming a fertile field of taxation. The following are the major trends in this regard: (1) Taking the field of motor vehicles taxation as a whole, Bihar, Mysore and Uttar Pradesh are the only states that have not substantially resorted to it. (2) Diversification or enhancement in motor spirits taxation. Uttar Pradesh and Rajasthan brought diesel oil within the purview of this taxation, following the example of other states. Assam, Gujarat, Madhya Pradesh, Maharashtra and Orissa have stepped up the rates. The rates, of course, vary from state to state but the sales tax on motor spirit has now reached as much as 12 paise per litre in Gujarat. (3) Introduction, diversification or enhancement of the tax on passenger fares and freights.

Assam, Kerala, Orissa and Gujarat belong to the first two categories, while many other states have increased the rates. Here again there is an inter-state spiral. Maharashtra has levied a tax on goods carried by private carriers in addition to the more usual tax on goods carried by public carriers. (4) There were increases in motor vehicles tax proper in all states except Bihar, Mysore, Orissa and Uttar Pradesh. Madras has introduced some refinements in the motor vehicles tax by relating the tax substantially to the daily mileage of passenger buses and laden weight of public carriers.

As regards sales tax, we shall confine ourselves to the directions in which the revenue from this source is sought to be increased: (1) There is a greater consciousness of the need to check evasion as evidenced by the steps taken by Kerala, Rajasthan, Uttar Pradesh and Madras. Uttar Pradesh took credit for Rs. 1.5 crores for 'adjustments and tightening' of sales tax. (2) Increase in the sales tax on luxury goods to 10% in pursuance of the decision of the Finance Ministers' Conference in February, 1963. (3) Adjustments made in sales taxation to correspond to the inter-state rate of 2% (now 3%); for example, Madras, which increased the sales tax on cotton, cotton yarn, cotton waste, etc. to 2% and Gujarat. (4) Increasing tendency to tax food grains and cereals, as in Kerala, Madhya Pradesh, Orissa and Rajasthan. (5) Inclusion of the central government within the scope of definition of 'Dealer' for the purpose of levy of sales tax, in Maharashtra. (6) Though comparisons are difficult and require to be made cautiously, relatively high increases in rates of General Sales Tax in certain states. Punjab increased its general sales tax from 4% to 5% and then to 6%.

There is obviously need for tightening of evasion and for interflow of information and research on sales tax. The question also arises whether there is not need for an all-India survey of the structure of sales tax as well as the procedures in various states. This will throw up a wealth of material conducive to rationalization in individual states.

We may now briefly refer to the position in respect of other taxes. One feature is the increasing exploitation of state excise after a period of stagnation. This trend is, however, confined to a few states only. Andhra and Uttar Pradesh have proposed to get sizeable revenues from this source, the former by the reorganization and tightening of neera shops (Rs. 3 crores) and the latter by changes in vend fees and rate of duty on country spirit and rum (Rs. 1 crore). On the side of entertainment taxes, there have been marginal increases in many states, though it is not clear whether the limits of taxable capacity have been reached under this source.

The imposition of electricity duties or taxes on consumption of electricity is probably the most important development in state finances during these years. It has been largely resorted to though it is only in Madras that it seems designed to fetch considerable revenue. Such a tax deserves to be widely adopted, for it will be a growing revenue in the context of increasing generation of electricity. Moreover, carefully and differentially applied, it is not a burdensome form of taxation.

New forms of taxation adopted during this period are: (1) Tax on state trading in kendu leaves in Orissa; (2) Tax on certain forms of advertisement in Rajasthan; (3) Tax on purchase of cows and she-buffaloes for disposal outside the state in Rajasthan; and (4) a toll on new roads and bridges in Orissa. These are, of course, taxes of limited application and very limited potential.

We may refer briefly to the non-tax measures taken: (1) Imposition of fees in all classes above primary stage in Punjab. Kerala proposed a similar measure but withdrew it. Kerala, however, increased the admission and tuition fees in colleges. (2) State trading in kendu leaves in Madhya Pradesh and Orissa. (3) Increases in state transport fares in Kerala and Madras. (4) Rationalization of rates of royalty on timber in Assam and increase in rates of royalty on minerals in Madhya Pradesh and Rajasthan. Madhya Pradesh in particular seems to have taken up the question of royalties, grazing and mining rules, etc. in detail. The royalties on minerals would require to be looked into by all the states, for the mineral rules were framed long back and may not be quite appropriate now. (5) Madhya Pradesh proposed to convert 50% of the beds in the general wards attached to the district hospitals and medical colleges as paying beds and to levy a charge of Rs. 2 per day per bed for persons whose monthly income exceeds Rs. 500. This is a noteworthy feature of a state government being willing to levy a fee on an existing social amenity.

Looking at the taxation effort of the states as a whole over fifteen years, certain features stand out. Certain states have done well in the matter of additional taxation. If mention should be made, Madhya Pradesh, Rajasthan, Punjab, Maharashtra and Madras belong to this category. States like Bihar, West Bengal and Uttar Pradesh do not seem to have been eager to raise additional resources.

It will have to be remembered that there is necessarily no relation between the quantum of additional taxation actually realised by states and the extent to which they fulfil their plans. States may fulfil their Plans without fulfilling their target for additional taxation or they may not fulfil their Plans in spite of fulfilling the targets for additional taxation. Such circumstances may arise for a number of reasons.

The nature of the additional taxation effort points on the one hand to the limitations in the sphere of taxation of states and on the other to a number of possibilities. Not least, it points to the need for a careful study of the taxation structure of the states, so that concerted attempts at raising revenues may be made. A close study of the taxation system of each state with reference to the economic conditions and other features will surely suggest avenues of additional taxation. Though the decision to tax is ultimately political and its timing is also political, it should be possible for each state to work out for itself the ambits of its taxation and the manner in which it could be undertaken.

With proposals for abolition of the land revenue in the air, the future of state taxation is getting restricted indeed and it is necessary to consider the line of approach of the states to future taxation. To put it briefly, if land revenue is 'abolished' at the state level, it may be delegated to local bodies for exploitation. Motor vehicles taxes are no doubt good sources of revenue but the comparatively slow production rate of automobiles sets a limits to their buoyancy. Far more careful study and attention to sales tax is necessary. This can bring in more revenues even at existing rates if evasion is checked. Electricity duties bid fair to become a very important source of revenue. All these major lines of taxation require a good deal of study and concerted attempts at additional taxation are necessary.

Proposals for additional taxation have often to be linked with schemes of social amenities, to make them acceptable and this contingency contains in it the seeds of a vicious circle. This brings us to the question of the revenue expenditure of the states.

CHAPTER IX

TRENDS IN REVENUE EXPENDITURE

Ultimately it is the size and composition of public expenditure that the community looks to in return for the taxes paid by it. As is everywhere the case, much less attention has been paid to the aspects of state expenditure than to state taxes. We have already seen that various classifications of expenditure are possible and we need not go into them in detail here. In our attempt to see what the state governments have done with the taxpayer's money we shall use the concepts of development and non-development expenditure.

We may note at the outset that the classification of expenditure like 'development' and 'non-development' and 'plan' and 'non-plan' sometimes endows certain schemes with more than necessary legitimacy and is therefore apt to distort one's view of public expenditure. After all, the test of public expenditure is the amount of satisfaction it gives to the public by the quantity or quality of the services it makes possible. But classifications like the development and non-development expenditure may ignore this point unless the Finance Department of the state retains its sense of proportion. Otherwise non-development expenditure will be frowned upon even though such an expenditure may be quite necessary, whereas development expenditure will be favoured even if a particular scheme of development is not very urgent or useful. Thus the appointment of staff for say the survey and sub-division of landholdings may cause more satisfaction to the public than, say, the provision of a refrigerator for medicines in a veterinary dispensary. Still an item coming under non-development expenditure is likely to receive lesser priority even though it may be quite necessary and urgent. Similarly, development expenditure outside the plan is likely to be less favoured than plan expenditure. For example, the maintenance of buildings of schools or primary health centres or roads, created in the previous Plan will receive less attention than the formulation of new ones of the same type. Thus maintenance and conservation are likely to suffer whereas a plan scheme, even if it is not very important, acquires a priority and urgency out of proportion to its real worth simply because it is in the plan. This distorts not only a sense of objectivity in our appreciation of the different directions of expenditure but can give room for an unnecessary rush and 'overdo' of expenditure of plan schemes.

This of course does not mean that development expenditure should not have priority. It only means that while we ensure the fulfilment of the plan, the sense of proportion is not lost.[1]

We may now turn to the substantive aspects of our study of the trends in public expenditure. What are the overall trends in public expenditure in fifteen years of planning and what are the directions of change? What are the levels of expenditure in individual states overall and in respect of particular schemes? How far is the money that the state governments spend their own and how far is it that of the centre? How far is the money available to the state government spent by itself or through other agencies like local bodies? A number of important questions like these could be raised.

There are the usual difficulties in comparison of expenditure figures among states and over the years but thanks to a careful study made in the Reserve Bank of India *Bulletin*, June, 1966, it is possible to make a comparative study of the position.[2] Table 9.1 shows the trends in total expenditure and development expenditure in various states as in 1951–52 and 1965–66. This would also show indirectly the trends in non-development expenditure.

The broad conclusions from the table are: (1) the total expenditure of the states has risen nearly three and a half times (at current prices) in fifteen years of planning. Development expenditure has risen by nearly five times and non-development expenditure by nearly four times. In other words, though there has been a shift towards development expenditure in all the states, the shift has not been uniform and besides non-development expenditure has its own rate of growth. As we have already stressed, non-development expenditure cannot in all cases be reckoned as some kind of a second class expenditure. The reasons for growth of non-development expenditure have been the following: (1) The integration of and/or re-organization of states (this had revenue benefits also); (2) Par-

[1]The Fourth Plan requires that the rate of increase in the expenditure on administrative services and tax collection as well as the non-plan development expenditure be restricted to 5 per cent per year. (*Fourth Plan, A Draft Outline*, p. 81). It is not known how far this would materialise. A natural calamity or increase in emoluments of government servants may throw this expectation out of gear.

[2]Estimates for the year 1951–52 were worked out for the present states with reference to population. See the Bulletin for details.

TABLE 9.1

Trends in Revenue Expenditure

(Per capita, Rupees, current prices)

	1951–52			1965–66		
States	Total expen-diture	Develop-ment expen-diture	Per-centage of (3) to (2)	Total expen-diture	Develop-ment expen-diture	Per-centage of (6) to (5)
(1)	(2)	(3)	(4)	(5)	(6)	(7)
Andhra Pradesh	12·1	6·0	50	40·7	26·4	65
Assam	12·4	6·4	52	49·9	31·3	62
Bihar	8·5	4·5	54	22·0	12·9	58
Bombay	15·7	7·6	49	49·6	25·8	52
Kerala	14·1	7·2	51	43·5	30·7	71
Madhya Pradesh	8·7	4·2	48	32·6	20·4	62
Madras	11·0	6·3	57	45·5	28·1	62
Mysore	19·0	11·1	58	42·2	27·2	64
Orissa	7·4	4·2	56	46·5	28·0	60
Punjab	12·9	5·1	40	52·1	29·2	56
Rajasthan	9·8	4·0	40	38·0	22·2	58
Uttar Pradesh	8·8	4·1	47	28·7	16·2	56
West Bengal	14·2	6·3	45	42·2	23·4	55
All States	11·6	5·8	50	38·9	22·9	59

[Source: Tables 3 and 4, State Governments' Expenditure, 1951–52 to 1965–66, Reserve Bank of India Bulletin, June 1966.]

ticular circumstances like partition, famine, floods and emergency;[1] (3) the steady increase in the emoluments of government servants necessitated by rise in prices; (4) the increase necessary to meet the needs of planning and to raise the level of services; (5) not least is the growing tend of interest charges. The interest charges have grown phenomenally over the years and the average for the Third Plan was about fourteen times the average for the First Plan. In this respect the growth of non-development expenditure is itself related to the growth of development expenditure on the capital account.

[1]Since 1957–58, states have on an average spent over Rs.15 crores per annum on natural calamities. The highest average annual expenditure is that of West Bengal with Rs.5·6 crores and the lowest that of Kerala with Rs.8 lakhs. See Table 32 of Appendix to *Report of the Finance Commission*, 1965.

The maximum shift from non-development to development expenditure (by 20 points) was observed in Kerala. There are states like Bombay and Madras where the shift was only by 3 points and 5 points respectively.

The trend of increasing non-development expenditure including expenditure on general administration raises the question of the size of the states and their relation to non-development expenditure. Discussions on the size of states and their relation to economic development are usually riddled with 'Cartesian doubts' and agnostic conclusions. Table 9.2 on the growth of expenditure on administrative services is interesting. The highest is as much Rs. 8·1 in Assam whereas the lowest is Rs. 2·8 in Bihar. All states have had appreciable increases in the per capita expenditure on administrative services over the fifteen years, with the exception of Bihar.

TABLE 9.2

Expenditure on Administrative Services
(Per capita, Rupees, current prices)

States	1951–52	1965–66
Andhra Pradesh	3·4	5·4
Assam	2·6	8·1
Bihar	2·2	2·8
Bombay	4·2	6·6
Kerala	2·3	4·4
Madhya Pradesh	2·4	4·3
Madras	3·1	6·0
Mysore	3·6	4·6
Orissa	2·1	5·0
Punjab	4·0	7·4
Rajasthan	2·9	5·0
Uttar Pradesh	2·3	4·3
West Bengal	3·9	6·7
All States	3·0	5·2

[Source: Table 5. State Governments' Expenditure 1951–52 to 1965–66, Reserve Bank of India Bulletin, June 1966.]

We may now proceed to analyze the directions of change in expenditure. Table 9.3 shows the trends in per capita expenditure with reference to the development on human and physical capital. Expenditure on education, medical and public health is classified as expenditure on human capital and that on other items like agriculture, irrigation, electricity schemes etc. as expenditure on physical capital. It will be seen that while human capital for all the states increased by as much as 4·2 times the physical capital increased by only 3·8 times. There has been a slight shift from the accent on physical capital to accent on human capital. This is naturally due to the emphasis placed on schemes of education and public health.

TABLE 9.3

Development of Human and Physical Capital
(Per capita, Rupees, current prices)

States	1951–52 Human Capital	1951–52 Physical Capital	1965–66 Human Capital	1965–66 Physical Capital
Andhra Pradesh	2·8	3·2	11·4	15·0
Assam	2·9	3·5	12·6	18·6
Bihar	1·4	3·1	5·9	7·0
Bombay	3·8	3·8	12·1	13·7
Kerala	3·7	3·5	20·2	10·5
Madhya Pradesh	1·9	2·3	11·3	9·1
Madras	2·8	3·5	14·2	18·9
Mysore	4·3	6·8	13·0	14·2
Orissa	1·4	2·8	9·4	18·6
Punjab	2·2	2·9	12·7	16·5
Rajasthan	2·3	1·7	12·2	10·0
Uttar Pradesh	1·7	2·4	7·8	8·4
West Bengal	2·9	3·4	11·0	12·4
All States	2·6	3·2	11·0	11·9

[Source: Table 3, State Governments' Expenditure, 1951–52 to 1965–66. Reserve Bank of India Bulletin, June 1966.]

Note: For definition of Human and Physical Capital, see text.

As regards states *inter se* there have been significant variations in total expenditure per capita between 1951–52 and 1965–66. In 1951–52 the level of per capita expenditure was the highest in Mysore at Rs. 19 followed by the erstwhile Bombay (Rs. 15·7), West Bengal (Rs. 14·2) and Kerala (Rs. 14·1); it was the lowest at Rs. 7·4 in Orissa. During the period of fifteen years the average per capita expenditure level for all the states increased more than threefold from Rs. 11·6 in 1951–52 to Rs. 38·9 in 1965–66. Rates of increase have been marked in states like Orissa (sixfold), Assam,

Madras, Punjab and Rajasthan (fourfold) but a small increase (about 2½ times) was observed in Bihar and Mysore. As a result the inter-state per capita levels underwent significant changes over the period. For instance, in 1965–66, Punjab, Assam and Orissa ranked first, second and fourth respectively in the level of per capita expenditure as against fifth, sixth and thirteenth respectively in 1951–52. Mysore which was the first and West Bengal the third in 1951–52 went down the list to the seventh and eighth position respectively in 1965–66.

Table 9.4 shows the per capita expenditure of various states on important items.

TABLE 9.4

Per Capita Expenditure on Important Items

(Rs., current prices)

	Education		Medical and Public Health		Agriculture and Veterinary	
States	1951–52	1965–66	1951–52	1965–66	1951–52	1965–66
Andhra Pradesh	2·0	8·1	0·8	3·3	0·7	3·6
Assam	2·0	8·9	0·9	3·7	0·9	3·7
Bihar	0·9	3·9	0·5	2·0	0·6	2·0
Bombay	2·8	8·4	1·0	3·7	0·9	3·4
Kerala	2·5	15·5	1·2	3·7	0·9	3·4
Madhya Pradesh	1·3	8·7	0·6	2·6	0·5	2·3
Madras	1·9	10·5	0·9	3·7	0·8	3·7
Mysore	3·1	9·8	1·2	3·2	0·9	3·4
Orissa	0·9	5·9	0·5	3·5	0·5	5·1
Punjab	1·5	9·0	0·7	3·7	0·8	4·3
Rajasthan	1·4	8·0	0·9	4·2	0·2	2·9
Uttar Pradesh	1·2	5·5	0·5	2·3	0·8	2·0
West Bengal	1·3	7·3	1·6	3·7	1·0	4·0
All States	1·7	7·8	0·8	3·2	0·7	3·1

[Source: Table 5, State Governments' Expenditure, 1951–52 to 1965–66, Reserve Bank of India Bulletin, June 1966.]

The expenditure on education rose from 15% of the total expenditure of states to 20%. Even now there are significant variations in the levels of per capita education expenditure in various states. The per capita expenditure in Kerala which is the highest (15.5) is four times as much as the per capita expenditure in Bihar (3.9). Number of states like West Bengal, Uttar Pradesh, Orissa and Bihar have levels of their per capita expenditure on education less than the all India Average. The wide disparities in educational expenditure underline the need for further planning in this regard. The emphasis of different states on University, Secondary and Primary Education will be seen from Table 9.5.

TABLE 9.5

Education Expenditure of States in 1960–61

(Rs. Crores)

States	University	Secondary	Primary	Total
Andhra Pradesh	1·66	4·38	8·55	17·38
Assam	0·73	1·96	2·51	7·10
Bihar	1·57	1·59	7·36	13·26
Gujarat	0·89	1·80	7·01	10·87
Kerala	0·86	2·93	10·88	16·16
Madhya Pradesh	2·10	3·46	6·23	14·33
Madras	1·30	3·83	10·14	18·39
Maharashtra	1·70	5·56	12·76	23·44
Mysore	1·15	1·44	7·11	12·47
Orissa	0·50	0·89	1·76	4·33
Punjab	1·14	4·77	3·82	11·28
Rajasthan	1·39	4·10	3·21	10·51
Uttar Pradesh	1·25	3·50	6·03	17·82
West Bengal	0·89	1·55	0·78	16·76
Jammu and Kashmir	0·21	0·91	0·45	1·89

[Source: Combined Finance and Revenue Accounts of Central and State Government in India for 1960–61, pp. 282–3.]

Note: The total will be more than the sum total of university, secondary and primary since certain other expenditures are also booked under the education head of account.

On the other hand the expenditure on Medical and Public Health does not reveal wide disparities and the differences have been small in 1951–52 as well as 1965–66. Even here Bihar has the lowest and Kerala the highest level of per capita expenditure.

Since the commencement of planning in India, the state expenditure in absolute terms has increased fourfold over the three Plan periods and nearly double in relation to national income from 4·1% to 8·2% in 1964–65. Over this period there were also substantial changes in the relative per capita expenditure levels in the states as also the pattern of expenditure. Table 9.6 shows clearly the shifting pattern of the allocation of the tax-payer's rupee by the states as a whole.

TABLE 9.6

Allocation of Tax-payer's Rupee by State Governments

(In Paise)

Development	1951–52	1965–66
Education	14·7	20·0
Medical and Public Health	7·1	8·2
Agriculture, Veterinary and Cooperation	6·3	8·0
Rural Development and Community Development Projects	0·3	4·9
Irrigation	4·1	3·1
Electricity Schemes	1·1	0·2
Civil Works	10·0	6·8
Industries and Supplies	2·6	1·7
Others	3·5	5·8
Non-Development		
Direct Demands	8·7	4·3
Debt Services	2·2	14·0
Civil Administration	26·1	13·5
Famine	0·9	0·8
Miscellaneous	9·7	5·6
Others	2·6	3·1
TOTAL	100·0	100·0

[Source: Table 6, State Governments' Expenditure, 1951–52 to 1965–66, Reserve Bank of India Bulletin, June 1966.]

We now turn to the question as to what portion of their expenditure the states have been able to meet out of their own resources. In this regard we have already seen that states have had very little surpluses after meeting their non-development expenditure. The surplus over non-development expenditure has not grown markedly during the plan periods, in spite of the buoyancy of state revenues and the efforts of additional taxation, the increase in revenues having been absorbed particularly by the mounting interest charges on debts. In fact the annual interest charges on debts are more than Rs. 120 crores which is one fifth of the additional taxation target for the Third Plan. This is a very strange situation indeed! This shows how much the growing predominance of the capital account has affected the state finances.

It will be interesting to see in 'per capita language' the part played by central assistance in the increase of per capita expenditure of the states from Rs. 11·6 in 1951–52 to Rs. 38·9 in 1965–66. Central assistance including shared taxes was Rs. 2·46 per capita in 1951–52 and was Rs. 13·69 per capita in 1965–66. Of the increase in per capita expenditure during 15 years more than 40% can be said to have been contributed by increased central assistance.

It is necessary to note briefly some trends in the expenditure policy of the state governments. One noteworthy trend has been the uniform keenness of state governments to provide social amenities. The provision of drinking water supply and village roads has received great impetus in recent years. Apart from aiming at universal primary education state governments have had special schemes in the education front like the school meals programme in Madras and also the pre-school programme. Madras has, besides, made education free up to secondary education stage. States have also interested themselves in some social assistance measures. States like Madras, Utter Pradesh, Andhra and Kerala have started old age pension schemes during the Third Plan for the help of old destitutes. The commitments on this score have mounted up in some states. There has been some attention to service conditions of employees and liberalized pension schemes and family pension schemes have been introduced in a number of states. State governments have often found it necessary to initiate new schemes of social amenities of appeal to the public in order to justify their tax measures. As we have pointed out already, there is an element of the vicious circle in tying up unbudgeted amenities with taxes to cover budget deficits.

Another trend in state expenditure has been that an increasing but not always substantial portion of the expenditure is incurred through local bodies. Figures are not available for bringing this out clearly. Mention of this will be made in the chapter on 'State-Local relations'. This is a trend worthy of note. In some states like Maharashtra, the percentage of grants to local bodies in total revenue expenditure is considerable indeed.

The question of economy in government expenditure is a vital one though it is often approached in rather naive terms. Attempts at economy are nothing new and have been made by states as early as 1949–50. In recent years there has been an increasing emphasis on economy measures and the Fourth Plan specifically provides for Rs. 250 crores as savings from economies (i.e. Rs. 59 crores per annum or, say, Rs. 4 crores per state). We have already seen how non-development expenditure is sometimes cut arbitrarily for the sake of economy and for the sake of plan schemes. But there is no doubt that there are pockets of government expenditure where economy is feasible and the real problem is to identify those pockets.

As an illustration of the measures on economy attempted by state governments we may give in Table 9.7 economy measures approximating to Rs. 1 crore per annum undertaken by the Government of Madras in 1966 in pursuance of the recommendations of its High Power Committee on Economy and Administrative Reorganization.

TABLE 9.7

Economy Measures in Madras State in 1965–66

(Rs. Lakhs)

	Annual Savings
Economy in Travelling Allowances and office contingencies	25·00
Revision of yardsticks for appointment of peons	20·00
Reduction in non-technical establishments	30·00
Winding up of Fuel Stores in Madras city	0·85
Abolition of the Department of Weights and Measures and merger with Labour Department	5·00
Reorganization of the Agricultural Income Tax Department*	6·00
Reorganization of the Registration Department†	9·50
Economies in Community Department Staff pattern—reduction in the number of Extension Officers (Industries) and down-grading of the posts of Extension Officers (Cooperation)	8·00
Introduction of officer-oriented system in the Finance Department	0·65
TOTAL	105·00

*The work will be done in the non-plantation areas largely by the Revenue Department.

†Documents used to be copied in the books of the Registry office. Now a copy produced by the party will be filed.

[Source: p. 14, High Power Committee on Economy and Administrative Reorganisation (Madras State), a Preliminary Report, 1966]

There are only two further comments necessary at the attempts on economy made by state governments. Firstly, economy should not be at expense of efficiency. A percentage cut in the staff will no doubt have a salutary effect and make people sit up and take notice. At the same time unless attempts are made to match economy with an increase in efficiency there will be no overall improvement. A mere mandate to cut travelling allowance or contingent expenses has no doubt a sobering effect on extravagant officials but is not the only or proper method of approaching the whole question. What is needed is the vigorous and assiduous scrutiny of all existing items of government expenditure including plan schemes so that the pockets of disguised extravagance are identified. As was pointed out by a committee of the House of Commons there is a real danger of a scheme living on its legend and going on because it was approved four or five years ago, without anyone saying 'why are we doing this?'[1]

We have so far been looking at government expenditure merely in terms of certain heads of services related to the budgetary form of state governments. An economic classification of the same expenditure in terms of economically significant categories is admittedly necessary. While the Government of India publishes an 'Economic Classification' of its budget along with other budget documents, state governments have been unable to do so. Some state governments like Madras have occasionally attempted such economic classifications but there has been no steady and systematic effort of this kind. There have however been a few *ad hoc* studies[2] and we have to rely on them for getting an idea of the economically significant categories of state government expenditure. Table 9.8 shows the economic classification of expenditure of state governments in 1958–59. It will be seen that wages and salaries form 42% of the total expenditure and commodities and services another 20%.[3]

One more point is necessary to complete the perspective. It would have been noticed that our discussion of the expenditure of state governments has proceeded in terms of current prices. Since 1951–52 prices have increased substantially and hence the figures of government expenditure in real terms will be less than what they

[1]p. ix, ***Treasury Control of Expenditure***, Sixth Report from the Select Committee on Estimates 1957–58, HMSO.

[2]Punjab University for 1958–59 and National Council of Applied Economic Research.

[3]These are exclusive of wage/commodity components in grant expenditure.

TABLE 9.8

Expenditure on Current Account of Government Administration and Commercial Undertakings for all States

(B.E. 1958–59, Rs. Crores)

	Government Administration	Commercial Undertakings
1. Consumption Expenditure	426	80
1.1 Wages and Salaries	297	47
1.2 Commodities and Services	129	33
2. Transfer Payments		
2.1 Interest	25	32
2.2 Grants to		
(a) local bodies	41	
(b) cooperative institutions	10	
(c) educational institutions	45	
(d) rest	18	
2.3 Subsidies for		
(a) agricultural purposes	2	
(b) non-agricultural purposes	2	
2.4 Pension or pension payments	15	2
2.5 Other current transfers	24	—
3. Provision for depreciation	—	5
TOTAL	608	119

[Source: Pp. 22–24, Central and State Government Budgets in India, an Economic Classification, 1958–59, Rangnekar and others, Punjab University, Chandigarh.]

are in money terms. The difficulty about deflating the 1965–66 expenditure arises from the lack of a suitable index, as no index has been constructed with this object in mind. The common indices like the wholesale prices index and the consumer price index will not obviously be applicable. Since a considerable portion of government expenditure is on wages and salaries one would have to construct an approximate index of the increase in wages and salaries. If one could hazard a guess keeping in mind the trends of prices, the trends in the emoluments of government servants etc. the 1965–66 figures may have to be deflated by roughly 30%. This would mean that the real rate of growth of public expenditure of state government is not fourfold but somewhat less than threefold.

CHAPTER X

CAPITAL EXPENDITURE

We shall now deal with the capital expenditure of state governments. In an era of planning and development, capital expenditure assumes a great deal of importance in terms of sheer magnitude alone. It has also got its economic effects depending on whether the projects financed by capital expenditure are quick yielding or long yielding in economic benefits. It has besides its implications in terms of its impact on the revenue budgets of the states. A well-thought out policy of capital expenditure forms the base of economic development in a country like India.

The general principle of classification of capital expenditure in India is that it is the kind of expenditure intended for creating concrete assets of a material character or of reducing recurring liabilities like pension payments. Capital expenditure can be financed by the Revenue Account or outside it. We have seen already that within limits there is a certain amount of discretion for the state governments for the classification of capital expenditure and that such discretion was exercised in 1955. State governments usually decide that a certain part of capital expenditure will be met from revenue, particularly works of less than a certain value. The rest alone are treated as capital expenditure outside the revenue account.[1]

We shall first of all consider the manner in which state governments obtained resources for financing capital expenditure and the directions in which they spend such capital resources. Table 10.1 will show the capital receipts of the states in the first three Plans.

The first thing that strikes one on seeing the table is of course the fourfold increase in the capital receipts of the states between the First and the Third Plan. The increase has been particularly marked between the Second and the Third Plans.

The next thing that strikes one is the phenomenal importance of the loans from the centre in the capital structure of the states. Nearly two-thirds of the capital receipts accrue from loans from the centre. The next important source is the open market loans

[1]See also Chapter III.

from the public. As will be explained in a separate chapter, apparently unimportant items like unfunded debts, deposits etc. also add to the capital receipts of the states.

It is worthwhile at this moment to remember that a considerable portion of capital expenditure is also directed towards the repayment of loans to the centre or to the public. Thus, in the Third Plan, as against a capital receipt of Rs. 449·3 crores from the public Rs. 123·1 crores had to be spent in the discharge of previous debts. Similarly, as against a capital receipt of Rs. 3,091·4 crores from the centre Rs. 1012·8 crores had to be spent by way of repayment. Repayment problems will arise in much greater magnitude in the ensuing plans.

TABLE 10.1

Capital Receipts of States

(Rs. Crores)

	1951–52	First Plan	Second Plan	Third Plan	1966–67
Permanent Debt	11·8	157·9	337·4	449·3	92·7
Floating Debt	3·7	15·7	28·9	17·2	– 19·0
Loans from the centre	74·0	769·5	1,417·0	3,091·4	626·8
Other Loans*	—	—	54·6	201·2	41·9
Unfunded Debt	2·6	26·9	49·9	99·2	30·3
Loans and advances repaid to state governments	24·3	122·9	231·7	386·8	124·4
Deposits and advances and other items	18·6	21·6	122·4	383·7	112·1
Total receipts	135·0	1,114·5	2,241·9	4,628·8	1,009·2

*Includes loans from National Agricultural Credit (Long-term Operations) Fund of the Reserve Bank of India, loans from National Cooperative Development Corporation and Central Warehousing Corporation, loans from Khadi and Village Industries' Commission, Employees' State Insurance Corporation, etc.

1966–67 figures are Budget Estimates.

[Source: Finances of State Governments, Reserve Bank of India Bulletin, May, 1966.]

We now come to the directions in which capital expenditure has been incurred. Table 10.2 will show the disbursements in the capital budgets of the states.

TABLE 10.2

Capital Disbursements of States

(Rs. Crores)

	1951–52	First Plan	Second Plan	Third Plan	1966–67
A. Capital Receipts	135·0	1,114·5	2,241·9	4,628·8	1,009·2
B. Disbursements					
Development outlay	100·3	698·8	1,260·0	1,811·5	399·5
Multi-purpose River Valley Schemes	27·6	235·1	262·9	290·7	51·8
Irrigation and Navigation	24·0	182·7	321·7	554·5	111·4
Schemes of Agricultural Improvement and Research	0·1	8·6	19·0	47·5	16·5
Electricity Schemes	19·9	134·2	166·2	107·8	32·9
Road Transport	1·6	9·1	17·3	24·8	−0·7
Buildings, Roads and Water Works	21·1	101·3	398·0	620·4	131·2
Industrial Development	5·8	26·9	68·8	148·9	50·5
Others	0·2	1·7	6·1	16·9	5·9
Non-Development outlay	27·3	−28·4	80·7	151·9	28·1
Other Disbursements					
Discharge of Permanent Debt	1·0	29·8	50·1	123·1	13·0
Repayment of Loans to the Centre	11·4	72·1	365·2	1,012·8	232·3
Repayment of other loans	—	—	5·7	44·9	32·0
Loans and advances by State Governments	48·7	291·6	610·0	1,577·3	370·3
Total Disbursements	188·7	1,063·9	2,371·7	4,721·4	1,075·0
C. Surplus or Deficit on Capital Account	−53·7	50·6	−129·8	−92·6	−65·8

1966–67 figures are Budget Estimates.

[Source: Finances of State Governments, Reserve Bank of India Bulletin, May 1966.]

One important point to note before any comparison is made is that the formation of Electricity Boards in various states will affect the comparison under electricity schemes. So long as the electricity undertakings were treated as departments of government the capital expenditure on electricity schemes would be shown as such in the budget. But once an Electricity Board is formed by statute the expenditure will be in the form of loans to Electricity Boards which will in turn incur the expenditure on capital.[1] This accounts for the very low figures shown as capital expenditure on the electricity schemes and partly accounts for the fivefold increase in loans and advances by state governments between the first and the third Plan.

Leaving electricity aside, it will be seen that the major items of capital expenditure have been towards multi-purpose river valley schemes and irrigation schemes on the one hand and schemes like buildings, roads and water works on the other. Financial stringencies have forced the state governments to defer all demands for capital expenditure outside the Plan, to such an extent that even essential schemes of a non-plan nature are perpetually deferred for 'better times'. For example, even the construction of latrines in taluk offices, the need for which requires no explanation and the amount required for which is very little, will be postponed in favour of some schemes of Plan expenditure whether they are equally urgent or not. The feeling that every expenditure shown as plan expenditure will earn central assistance puts an additional spur on the part of state governments to shut their eyes to all capital expenditure of a non-plan nature.

Reverting to the directions of the capital expenditure, it may be stated that by and large the substantial capital expenditure incurred, on irrigation and electricity schemes particularly, has built a good economic base of the country. But this does not mean that it is necessarily remunerative now in the financial sense. Not only has a portion of this expenditure been unable to produce profits which can augment current revenues of the state governments but they are not even able to pay the interest charges on the loan amount. The financial working of the Electricity Boards is being discussed in a separate chapter. But Table 10.3 will show the financial working of the irrigation and departmental electricity schemes for the First and Second Plan periods.

[1]However, some states like Andhra, Mysore and Orissa have retained the construction of power generation schemes in their hands. See Report of the Committee on the Working of State Electricity Boards.

TABLE 10·3

Financial Results of Irrigation (Commercial) Works and Electricity Schemes, 1951–52 to 1960–61

(Annual Averages, Rs. Lakhs)

State	Irrigation	Electricity	Total
Andhra‡	−229	−140	−369
Assam	..	1ˣ	1
Bihar	− 37	− 14	− 51
Bombay*	−132	− 10	−142
Madhya Pradesh	..	1*	1
Madras	−157	+ 16	−141
Orissa	− 22	+ 6	− 16
Punjab	+ 86	+ 35	+121
Uttar Pradesh	− 70	− 20	− 90
West Bengal	− 28	− 1	− 29
Kerala	− 54	..	− 54
Jammu and Kashmir	..	..	..
Hyderabad†	− 10	− 3	− 13
Madhya Bharat	+ 12	+ 7	+ 19
Mysore	− 61	+ 23	− 38
Pepsu†	+ 30	+ 9	+ 39
Rajasthan	+ 4	− 8	− 4
Saurashtra†	− 15	− 2	− 17
Travancore and Cochin†	− 3	− 14	− 17

*For both Maharashtra and Gujarat.

†Upto 1955–56 only. ‡From 1953–54. ˣFrom 1957–58 only.

[Source: Compiled from pp. 30–31 of *Report of Finance Commission*, 1957 and pp. 113 and 115 of *Report of Finance Commission*, 1961.]

Note: Averages are for 4 years ending March, 1956 and 4 years ending March, 1961 except irrigation schemes which include 1951–52 figures also.

The implications of this state of affairs are manifold. The question would arise whether the indirect subsidy to irrigation and electricity is a conscious choice or is because of defective planning to the extent that capital expenditure even in fields which can be expected to be productive is non-productive. The state government has no means of discharging its interest liabilities except by additional taxation. We are referring more to this aspect in the chapter on 'debts and debt servicing'.

The answer to the issue is not certainly that capital expenditure which cannot pay itself back should not be incurred. One has to take into account the indirect benefits and contributions of these capital projects to the economy rather than think only of a strict financial return. Many projects especially on the food front have to be undertaken to increase food production and when food production is such a dire necessity for the country an indirect subsidy would be worthwhile and inescapable. Hence the real question is not one of doing away with unproductive capital expenditure altogether but rather one of a conscious allocation of capital resources on productive and unproductive items having in mind how far the taxpayer should be burdened with interest payments. The moral of the capital budget is really to be seen from the revenue budget.

The poor returns from commercial irrigation undertakings are partly due to the unwillingness to levy suitable water rate and partly due to the lack of utilization of the irrigation potential by the cultivators.[1] State governments have not also been uniformly keen to levy betterment levies in respect of irrigation projects. The Third Finance Commission pointed out that in a state 'in real need of resources, the collection of betterment levy already introduced had to be suspended just because the neighbouring state had done so in a more prosperous contiguous area'.[2]

We have already referred to the undue importance that capital expenditure on the plan account has acquired. Ambitious schemes of a capital nature are liable to be undertaken by departments as they have a spectacular appeal. Economies in construction are very important and cost consciousness is imperative. Too many capital projects out of proportion to the annual availabilities of capital resources may also be undertaken with the result that the construction is spread over a number of years involving a delay in the completion of projects. This warrants a close look into the capital budgets of the state *vis-à-vis* their commitments on capital account.

There is one other aspect of capital expenditure which may be mentioned here i.e. the quantum of loans and advances issued to the cultivators and others by the governments. The grants of loans and advances to cultivators is quite necessary but the collection of such loans has not been as easy as the disbursement.[3]

[1]The utilization at the end of the First and Second Plans was only 48% and 71% respectively. At the end of the Third Plan the anticipated utilization is 77%. This is with reference to gross area actually benefited. See p. 216, Fourth Five Year Plan, A Draft Outline, 1966.

[2]Page 39, *Report of the Finance Commission*, 1961.

[3]See Table 11.2 in Chapter XI for the quantum of outstanding loans and advances by state governments.

A mention should be made at this stage of the fact that state governments incur a certain amount of capital expenditure on road transport services which are run either as departmental undertakings or as corporations. We give in Table 10.4 the financial results of these undertakings and their fleet strength but it may as well be mentioned at the outset that these undertakings may not in any case show any large contribution to general resources, in view of claims for higher wages on the part of the workers and of income-tax which these corporations will have to pay.

TABLE 10.4

Working of State Transport Undertakings (for 1962–63)

(Rs. Lakhs)

State	Fleet Strength of Buses	Gross Revenue	Total Operating Cost	Net Revenue
Andhra Pradesh	1,738	791	749	42
Assam	426	171	162	9
Bihar	882	319	287	32
Gujarat	2,665	1,085	1,080	5
Jammu and Kashmir	198	178	134	44
Kerala	839	433	394	39
Madhya Pradesh	830	222	221	1
Madras	814	406	378	27
Maharashtra	4,064	2,454	2,233	236
Mysore	1,962	836	803	33
Orissa	708	238	186	52
Punjab	1,070	506	339	167
Rajasthan	257	90	62	28
Uttar Pradesh	3,312	1,280	1,056	224
West Bengal	831	517	513	4

[Adapted from Statistical Abstract of the Indian Union, 1965, p. 311.]

There are other aspects of the loan policies related to capital expenditure and these we shall consider in a separate chapter.

CHAPTER XI

PROBLEMS OF DEBT AND DEBT SERVICING

We have had more than one occasion in the previous chapters to refer to the mounting volume of the debts of state governments. That the debts of state governments have increased considerably need not occasion any surprise since the effort of planned development necessarily involves a substantial volume of capital outlay. But there are three aspects of the situation which would cause concern namely, (1) the sheer volume of debt; (2) the mounting interest charges; and (3) the control exercised by the Centre through its lending and its control of open market borrowing.

An idea of the growth of debt during the three Plans can be had from Table 11.1 which gives the figures statewise.

TABLE 11.1

Growth of Indebtedness of States (as on March 31st)

(Rs. Crores)

State	1952	1957	1961	1966
Andhra Pradesh	..	125	228	439
Assam	4	26	54	115*
Bihar	12	76	187	418
Gujarat	..	..	114	259
Jammu & Kashmir	6	24	39	96
Kerala	15	44	89	203
Madhya Pradesh	23	57	145	347
Madras	50	109	201	403
Maharashtra	56	189	265	503
Mysore	21	75	138	227*
Orissa	20	91	160	225*
Punjab	63	202	268	407
Rajasthan	15	64	158	328
Uttar Pradesh	58	222	392	670
West Bengal	43	190	310	527
TOTAL	386	1,494	2,748	5,167

*1964 figures.

[Source: *Report of the Finance Commission*, 1965, pp. 212–4.]

It will be seen that in fifteen years the debt has on an average increased by over thirteen times. The increase of debt has varied from State to State. In some states it has increased by as much as thirty times the level of debt in 1952 and the lowest rate of growth has been eight times. The sheer volume of the debt, particularly to the Centre, and its complexity have been the subject matter of concern to the state governments, the central government and the Finance Commissions.

As regards the composition of the debt, Table 11.2 will show that it is the loans from the Centre that go to make up over 75% of the indebtedness of the states.

When the Second Finance Commission investigated the question in 1956, the outstanding debt of the states to the centre was only about Rs. 900 crores. Even then there was no uniformity in the

TABLE 11.2

Outstanding Debt of States (as on 31st March, 1966)

(Rs. Crores)

State	Loans from Centre	Permanent Debt	Total (including others)	Rev. Expr. on Debt Services	Debt due to States
Andhra Pradesh	338	80	439	17	163
Assam*	99	6	115	7	52
Bihar	342	48	418	21	116
Gujarat	192	54	259	18	91
Jammu and Kashmir	88	—	96	3	8
Kerala	145	35	203	3	96
Madhya Pradesh	293	36	347	15	179
Madras	272	95	403	17	214*
Maharashtra	335	115	503	26	170*
Mysore*	155	41	227	14	84
Orissa*	193	28	225	16	14
Punjab	368	6	407	22	127
Rajasthan	254	34	328	15	90
Uttar Pradesh	468	159	670	34	330
West Bengal	436	66	527	22	147
TOTAL	3,978	803	5,167	250	1,881

*Figures relate to 1964.

[Source: *Report of the Finance Commission*, 1965, pp. 205–8, 212–17.]

rates of interest charged or in the repayment periods. The rates of interest varied from 1 to 5% and the repayment period from 1 to 40 years. After detailed study the Second Finance Commission felt that the wide variations of the rates of interest and repayment periods had introduced avoidable complications in the financial relations between the Union and the States. It thought that it would simplify matters and save a great deal of labour and accounting if these loans were consolidated and the rates of interest and the problems of repayment rationalized. The Commission suggested that loans carrying 3% of interest should be consolidated at that level and that for repayment purposes loans of this type should be divided into those repayable before 20 years and after 20 years. Similarly it suggested that loans carrying less than 3% rate of interest should be consolidated at $2\frac{1}{2}$%.

The recommendations relating to repayment were not accepted by the Government of India as they envisaged the repayment of the entire amount on the expiry of the term of the loan. This payment procedure suggested by the Commission would create problems both for the Centre and the states although these problems might have been avoided by provision for a sinking fund or for annual repayment. The general average rates of interest charged by the government were consolidated but the loans given out of the Special Development Fund had their own amortization schedules and they were continued as such, the repayment periods being based on the terms of repayment to the foreign lending agencies concerned.

It was contended before the Commission by states like Bombay, West Bengal, Madras, Uttar Pradesh and Assam that the average rate of interest charged on the loans extended to states had been appreciably in excess of the average cost of the Centres' own borrowings. It was particularly true in the case of loans extended from the Special Development Fund which comprises the loanable part of the PL 480 counterpart fund. Some of the other states requested that the loans or parts thereof should be converted into grants. They also pleaded that resources provided by deficit financing should be made available free of interest to the states.

The Third Finance Commission did not and was not expected to come to grips with the problem in its various aspects. It however noted the mounting interest liability and the non-productive nature of some of the schemes taken up and suggested that the position was far from satisfactory and required analysis and review.

The Fourth Finance Commission has dealt with the matter in

greater detail. Its terms of reference required the Commission to consider whether a fund could be created out of the excess, if any, over a limit to be specified by the Commission, of the net proceeds of estate duty on property other than agricultural land accruing to the states every year. The Commission found that the total collections of estate duties were only of the order of Rs. 7 crores per annum and that the creation of a sinking fund out of this was not likely to ease the burden of repayment. Besides there was no virtue in creating an artificial sinking fund when there was no real revenue surplus.

The Commission however took the opportunity to go into the question of borrowing by states in some detail. The Commission felt that amortization should be created by it as a legitimate part of revenue expenditure irrespective of the sources of revenue i.e. whether revenues collected by the states or accruing by devolution or grants. The Commission recommended that an early enquiry through a representative and expert body should be undertaken to decide the principles of a scheme of amortization of the public borrowing by the states.

The Commission rightly pointed out that by far the most important problem in regard to the amount involved and more complicated in regard to the underlying the policy was the problem of borrowing by the States from the Government of India. It felt that over a growing area of public expenditure the relationship between the centre and the states was developing into an 'unlimited partnership'. A survey of the soundness of the present system of inter-governmental borrowing was considered by it to be necessary as much as in the interests of the states as that of the Government of India.

Various suggestions have been made in regard to a better ordering of the borrowing by the states particularly from the centre. One of the suggestions is that there should be a Central Loan Council as in Australia. The suggestions for such a Council are as old as the Percy Committee (1932) which went into the federal financial question. Will the constitution of such a council solve the problem or merely shift it?

It appears that we cannot come to grips with this problem unless we keep a clear view of the objectives and outlines of the situation. Nobody may be against borrowing as such but it is its utilization that has to be considered carefully. Again the states will not feel an adequate sense of financial discipline unless they are called upon to account for their borrowings in various ways. The present situation where the states lean heavily on the centre for capital resources is apt to give them little sense of responsibility for the use of the funds. The centre on its part may not also feel an adequate sense of responsibility in so far as it is only passing on the money to the states whose lookout it will be to service the repayment. Again, the mere inducement of the availability of the loan may result in states taking up schemes which they may not otherwise take up if funds were more difficult to get.

Not only will the states have a tendency to lose the sense of of financial discipline but the centre by the very process gains much more control over the states in financial matters than is perhaps necessary or desirable or was originally envisaged. A situation where the states have a greater responsibility for borrowing money would not only reduce the central control but make them more alive to the need for productive utilization of capital funds.

Another aspect of the problem is of course the servicing of debts. The first question that will arise is whether the states cannot reduce their capital expenditure. More importantly, mere exhortation that the states should cut down 'unproductive'[1] capital expenditure will not take care of the fact that there are various directions in which states have to utilize their capital resources, some of which are clearly unproductive or semi-productive.

It is because of this and of the various considerations we have set forth above that it is necessary to have a composite solution to the problem. The mere creation of a Central Loan Council without providing a set-up for a reasonable solution of the problem will only result in transferring it from one forum to another. At the same time, it may not be possible to completely eliminate central loans. Mr K. Santhanam for example writes, 'I do not also think that the Union Government should directly make any loans to the states. They should be encouraged to borrow directly from the public as much as they can. For the balance, they should borrow from the Reserve Bank subject to limits fixed by the Central Government which should stand guarantee for such loans. A special wing should be created for the purpose and the loans should be on a business basis in the same way as World Bank. A corps of inspectors and advisers should be built up by this wing to ensure that loans are made only for productive purposes and to watch the progress of the projects. Existing loans may be transferred to that wing.

[1]We shall be using the terms 'productive' and 'unproductive' in respect of capital expenditure in the sense that capital expenditure which cannot directly pay itself back is 'unproductive'.

It may be conducive to better relations between the states and the Union if out of the existing loans, a part, say, at the rate of Rs. 50 per capita for each state is written off.'[1] A complete subservience to productive projects, which one should no doubt emphasize for the sake of spotlighting, may not help the country's capital development in as much as it has to be realized that there are legitimate avenues of capital outlay which are not strictly productive.

The composite approach which we advocate would therefore take note of the fact that loans are utilized by the states in various ways some of which are productive, some of which are semi-productive and some of which are non-productive. The states also give loans in their turn to the local bodies, the public and to the Electricity Boards.[2]

A rough analysis of the types of expenditure incurred in the Plans on the capital and loan accounts will clearly illustrate the position and also give an idea as to how far the plan expenditure financed out of loans can be considered to the productive. Figures for this kind of analysis are difficult to obtain and hence we have to proceed on a number of assumptions and on a rough basis. If one goes through the heads of development in the Plans, one could isolate broadly the items of capital outlay. (1) Irrigation and power (including flood control) are capital outlays par excellence. These outlays can be legitimately expected to pay their way, even though in practice they often do not. (2) 'Roads' is an item of capital outlay. Unless a specific provision is made earmarking a portion of the proceeds of motor vehicles taxes (and this is unlikely) they cannot be reckoned as fully 'productive' items of expenditure. (3) Under the heads 'industries and minerals' and 'village and small industries', there is a certain amount of revenue expenditure as well as not fully productive capital expenditure but we may still assume that they are all items of productive capital expenditure. This would be a liberal assumption in favour of increasing the percentage of productive capital expenditure. (4) Most of the expenditure on minor irrigation and soil conservation is also of a capital nature but they cannot be taken as directly paying their way. A portion the expenditure on soil conservation is also by way of loans to cultivators to be repaid. (5) Other items of capital expenditure will be

[1]P. 785, Federal Financial Relations in India in Indian Finance dated 12th November, 1966.

[2]Loans to Electricity Boards are in many states unrequited. See Chapter XII for the finances of Electricity Boards.

on buildings, equipment, housing etc. which will be either semi-productive or unproductive.

Thus it is only the expenditure on irrigation and power and industries that can or should be made to, directly pay its way. Table 11.3 shows the proportion of what can be regarded as productive and what certainly cannot be ranked so in the plan capital outlay of state governments in relation to the total indebtedness of states on account of the plan.

The table has had to be necessarily approximate but it will be a safe bet to assume that 25 to 30% of the capital outlay under the Plans of the states will not be able to produce adequate returns to pay the interest even with the most prudent financial management. Any scheme regarding loans to states should take note of this fact.

In the context of the need for such capital outlays and to serve the three objectives that we have outlined above, namely greater sense of financial discipline of states, lesser control by the centre and greater provision for payment of interest and amortization charges, it is necessary to have a scheme on the following lines: (1) Under the Plan programme there are schemes like construction of buildings and capital expenditure related to social services that cannot be expected to yield a commercial return. The capital expenditure on roads will also have to be put under this category since in view of the financial stringency it would not be worthwhile to expect the states to link a portion of motor vehicles tax to road development. Similarly, expenditure on housing whether for slum clearance or for middle income group etc. will also have to be put in this category. These loans being obtained in furtherance of plan objectives in which the centre is interested should be admitted into the non-productive fold. However, in order to instil a feeling that there could be no loan without interest a uniform rate of 3% interest may be levied on the loans. Loans of this kind should be consolidated and issued to the states every year. (2) Schemes which ought to be productive should be grouped together and should be administered by the Central Loan Council on strictly commercial principles. Under this category would come irrigation and electricity schemes and capital outlay on industrial and commercial undertakings. These loans may be given every year on an agreed rate of interest say 6%. The Central Loan Council will advance these loans and expect a payment of interest as well as capital regularly. A Loan Council of this kind will naturally have in it the representatives of the States, Planning Commission, Finance Ministry, Reserve Bank of India and so on. By this means, a substantial portion of the capi-

TABLE 11.3

Estimate of Productive Capital Outlay of States in Second and Third Plans

(Approximate) (Rs. Crores)

Plan (1)	Magnitude of Productive Capital Expenditure (2)				Total indebtedness on Plan Account (3)	Percentage of (2) to (3) (4)
	Irrigation	Power	Industries	Assumed Total		
Second Plan	345	419	98	1,000	1,400	71%
Third Plan	581	883	217	1,800	2,800	64%

Note: Column (3) includes total loans from centre other than for natural calamities, net public debt and share of small savings but does not include unfunded debt or miscellaneous capital receipts. It has been computed from multiple sources. Assumed total under column (2) is to allow for productive expenditure under other heads, if any.

[Source: Multiple sources.]

tal needs of the states can be taken out of central control and at the same time put on commercial lines.

The question will naturally be raised that experience has shown that many electricity undertakings and irrigation projects have not paid their way. The answer is that it is necessary that they should either pay their way or the states should be prepared to subsidize them in an open form. Loans to local bodies[1] and the public may also be financed from the Central Loan Council so as to make states feel a sense of urgency for their recovery. (3) Schemes of other capital nature may be met from borrowing from the open market by the states. These may be schemes which are not included in the Plan but on which the states may feel it necessary to spend. But they will have to necessarily justify such schemes before the open market. A certain amount, say, the present level of market borrowing by the states, may be earmarked for plan schemes and over that the states can borrow for other schemes if they want to and if necessary, subject to an upper limit. Any amount that they borrow in excess of the limits, they will have to repay on their own and cannot expect any central assistance. The interest charges on this score will also be left out of reckoning by the Finance Commission on both the revenue and expenditure side. The States can thus have an elbow room within the limits of responsibility.

A composite scheme like this will make the state governments better aware of the need to take effective steps for repayment. It will also make the centre aware that a part of the loans it is passing on to the states cannot be expected to be productive and may therefore have to be regulated in volume. Such an awareness need not necessarily be there now in the choice of plan schemes and endowing them with patterns of assistance.

Surely, however, there are two difficulties in this composite scheme. (1) The fixation of Plan assistance may be a little more complicated. At the time of Plan formulation as well as during the annual Plan discussions an agreed formula will have to be worked out as to the amounts that the states should have under each item. The Loan Council will have to work in close coordination with the Planning Commission and have the same line of approach to plan problems as the Planning Commission itself. Of course there will be no multiplication of rates of interest and repayment schedules under each scheme. The non-productive loans given directly by the

[1]That is, loans to enable states to lend to local bodies. Suggestions for financial institutions for direct lending to local bodies have been made in recent years. See Chapter XIII.

centre may have a period of repayment of, say, 30 years. The productive schemes may have, a period of repaying of, say, 20 years. Market borrowings will of course have a period of about 10 years as at present. (2) The other difficulty will be about what we should do about the present outstandings: whether they should be written off or whether they should be allocated in the way we have indicated. The best course would be to consolidate and standardize the percentage of interest and the period of repayment and continue them on the existing basis.

It will be possible to work out a scheme like this in detail; only a composite scheme will help us to come to grips with the problems of debt.

In the time to come too, there will be productive as well as unproductive outlays to be financed by capital expenditure. While a good deal of emphasis is no doubt necessary where an adequate return is possible and should be insisted upon, it would be wrong on that score to call a complete halt to other programmes. Governmental activity in a planned economy is as much, if not more, to provide capital assets of value to the community as a whole as to provide current services. To that extent, the community should be prepared to recognize that such assets will create additional tax burdens on the community for servicing the 'unproductive' debt. What could be and should be argued is however that the 'mix' of capital outlay should be such that it strikes a balance between the tax burden on the community, present and future, and the national needs. Hence, the difficulties experienced by state governments in facing mounting interest charges on unproductive debt are not necessarily the result of impecunious folly.

Having said this, we should hasten to emphasize the other side of the argument, namely, that states must make their productive projects like irrigation and power projects pay their way. That the situation in this regard is unsatisfactory has been explained already. A Committee of Ministers which went into the working of the State Electricity Boards has recommended that the rate of return on capital employed in electricity undertaking should be raised to 11% per annum on the basis of a planned programme. The financial working of the electricity boards is taken up in another chapter and it will suffice here to say that the pace of implementation of the recommendation has not been as fast as one would wish. Another Committee appointed to suggest ways and means of improving returns from irrigation projects recommended that irrigation rates should be fixed at 25 to 40% of the additional net benefit to the farmer from irrigated crop and keeping in view factors like rainfall, water requirements, yield and value of crop; where it was not possible to measure this net benefit, the Committee suggested that the rates should be 5 to 12% of the gross income of the irrigated crop. The Committee also suggested that compulsory water charges sufficient to cover at least the maintenance and operation costs of irrigation works, should be made applicable to the entire area served by irrigation projects irrespective of whether water is drawn by the cultivators or not. These recommendations still remain to be implemented. Prima facie it appears that more co-ordinated and concerted action at the initial stages of planning may have avoided the drift of lack of effective action in this regard which seems to be prevailing and which is getting more and more difficult to correct.

All our discussions above have obviously had as the background the problems of amortization. A number of states have had sinking funds and also made provisions for 'appropriation for reduction or avoidance of debt' but the contributions to the sinking funds have not been uniform either in volume or regularity. The corpus of the fund is only shown in the books but it does not exist (except where it has been invested in securities) and has been used up for the ways and means of the government. The inadequate amortization has been partly due to the financial difficulties of state governments and partly due to the view taken by the first three Finance Commissions that the provision for amortization of market borrowing is admissible only to an extent to which a state's revenue resources other than grants to which they are entitled under Art. 275 can bear the financial burden. But the Fourth Finance Commission included as legitimate items of revenue expenditure, the provision already included in the budgets of state governments on account of amortization of their market borrowings.

It does not appear that the problems of amortization have received adequate attention and the present arrangements do not face the problem squarely. We cannot do better than to quote the Fourth Finance Commission on the evolution of amortization attitudes of state governments:

'It is necessary to be clear about the purpose and nature of a loan before its service and amortization can be put on a sound basis. When borrowing was largely confined to meeting either distress expenditure or the provision of a public amenity no serious doubt existed as to the burden both of interest and repayment being borne by revenues over an appropriate number of years. The mere fact that certain items of expenditure resulted in the creation of durable assets did not alter the fact that the expenditure had to be finally met out of revenue. Only items of expenditure which created a productive asset, bringing in a net revenue which would pay for interest and repayment, could be classified as investments and kept out of the revenue budget. A number of items fell between the two clearly defined classes, as being partly a revenue amenity and partly a capital investment. The extent to which each such item constituted a burden on general revenues had to be determined by the extent of its being an amenity and not an investment. This in substance was the prevailing practice of State and Central Governments till the developmental functions of both the state and central governments became increasingly important.

In 1955 the Government of India advised the state governments that all expenditure on capital assets, that is durable or fixed though not necessarily productive or self-liquidating assets, should be held eligible for being serviced out of loans, and that the amortization of such loans need not be treated as a charge on revenue except to the extent that the state governments were bound to provide in accordance with any law or with any specific undertaking given in the case of any loan. In its initial stages this practice, which ran counter to the more discriminating policy of the earlier period of keeping out of the revenue budget only productive and self-liquidating items of capital expenditure, did not produce serious results, though it appears that some at least among the state governments had repeatedly urged the claims of the more orthodox policy. The diversion of large items of unproductive or inadequately productive capital expenditure from the revenue to capital budgets made it possible to show a balanced revenue budget and to go on balancing the capital budgets also by fresh borrowings. As the sources and purposes of borrowings were numerous and ever on the increase, no serious question about the soundness of the new system projected itself for some time on the attention of governments. But as the burden of central loans began to pile up and as the unproductive i.e. non-revenue earning nature of a large part of it e.g. education, health, protective irrigation, etc. became clear, the states felt that any further continuance of this policy by the Government of India could only mean that the centre would ultimately take care not only of the interest, but also of the repayment liability of the whole debt, at least of that part of it which was not clearly productive of a net return equal to these obligations. As we have noted above, more than one state has urged us to take this view of their indebtedness to the centre.

As recently as March of this year the classification of expenditure as between capital and revenue accounts has received attention from the Finance Ministry of the Government of India. While it is recognized that only clearly productive items of capital expenditure can be kept out of the revenue budget no definite provision has yet been made to ensure the observance of this salutary principle. Even when the general principle is accepted its application, or rather its re-application in a developmental pattern of expenditure, is bound to raise a number of difficult procedural and financial issues. It is only in the light of a thoroughgoing investigation of past commitments and of future borrowings that the exact impact of central loans on state budgets can be measured, and incorporated into the scheme of devolution and grants which it is the function of the Finance Commission to recommend.'[1]

We may now devote ourselves briefly to the question of open market borrowings by state governments. We have already seen that it is central loans rather than open market borrowings that largely finance the capital outlays of states. Tables 11.4 shows that the quantum of net open market borrowings and their proportion to their total indebtedness vary from state to state.

Broadly speaking, while all states have been successful in open market borrowing, some states have been more successful than others. The quantum of the loan as well as the terms of flotation are decided for each state by the Central Government and the Reserve Bank of India. States with organized capital markets, like Maharashtra, West Bengal and Madras have found it easier to raise loans and at a larger volume than states like Orissa and Assam. Each state raises its own loan. In 1963 there was a common loan programme and apportionment among the states, but this was found unsatisfactory and given up.

Table 11.4 also shows that any proposal to allow the states to borrow their entire requirements from the open market may not work well, not only because of the high level of their requirements, but because different states have different borrowing capacities. Our composite proposal for state borrowings will be better appreciated in this light.

Some of the electricity boards also float their loans in the open market, but their loans are second class securities compared to those of the states just as those of the states are reckoned as inferior

[1]Pp. 66–67, *Report of the Finance Commission*, 1965.

TABLE 11.4

Public Debt of States (Rs. Crores)

State	Net Borrowings between 1952 and 1966	Permanent Public Debt as on 31.3.66	Col. (3) as a percentage of total debt
(1)	(2)	(3)	(4)
Andhra Pradesh	80	80	18
Assam	6	6	5
Bihar	48	48	11
Gujarat	54*	54	21
Jammu and Kashmir	—	—	—
Kerala	30	35	17
Madhya Pradesh	30	36	10
Madras	73	95	24
Maharashtra	92†	115	23
Mysore	25	41	18
Orissa	28‡	28	12
Punjab	6	6	1
Rajasthan	34	34	10
Uttar Pradesh	143	159	24
West Bengal	64	66	20
TOTAL	713	803	

*From 1961. †From 1952. ‡Nil in 1952.

[Source: *Report of the Finance Commission*, 1965, pp. 212–4.]

to that of the centre. The securities of the electricity boards and the states are quoted below their par value. A study of the distribution of the public debt of state governments among various types of institutional investors will show that institutional investors largely finance these loans.

The Life Insurance Corporation of India has also begun to directly advance loans to states for projects like housing and to local bodies for water supply schemes. The Reserve Bank of India does similarly for certain schemes under co-operation.

What we have said above is enough to establish that there is an urgent need for clear thinking and action on the problems of the debts of state governments. The Fourth Finance Commission felt that 'the scrutiny, classification and treatment of accumulated indebtedness would need an elaborate, expert and representative deliberation. We are convinced that in the interest of financial soundness such an enquiry ought not to be delayed any further'.[1]

[1]P. 67, *Report of the Finance Commission*, 1965.

CHAPTER XII

A MISCELLANY

In this chapter we shall deal with some of the less known but interesting aspects of state finances which are not usually treated at any length in studies of public finance. It is proposed to be shown here that net deposits and certain miscellaneous items play a considerable part in the financing of the capital outlays of state governments. The growing role of small savings will also be discussed as also the role of the electricity boards which are the greatest users of capital resources in state finances. We shall finally make some general observations on problems of financial administration and budgetary reform.

The role played by deposits and miscellaneous capital receipts is not usually known or appreciated as they are considered to be marginal items that can be ignored. We have seen particularly that the public account contains many transactions of a miscellaneous nature. There are receipts as well as expenditure on this account and every year there is a net surplus or deficit. This is ordinarily used by the government for financing its outlay. The public account consists of a number of items, some of which involve constant accretion of receipts and some receipts and expenditure of a more or less equal magnitude. Some of these items can be estimated properly and others are not capable of being estimated equally well.

In the public account we have the 'unfunded debt' section which consists of the contributions to state provident funds by employees. Under this head the receipts are usually much more than the disbursements and hence a substantial amount of money is available for immediate use. Then comes a large category of items under 'Deposits and advances'. The State Electricity Boards usually keep their accounts under this head of public account with the government. Any surplus in their accounts is of course reckoned by the Electricity Boards themselves for their capital programmes. We then have a number of funds like sinking fund, famine relief fund, sugar cane fund etc. The corpus of these funds may be large as contributions to them accrue year after year. But the corpus is invisible as the money is not kept intact but used for normal purposes of government. The depreciation reserve funds of commercial and non-commercial undertakings are also exhibited under this account. Then we have the deposits of local funds. Panchayati raj institutions and municipalities usually keep their accounts as personal deposit accounts in the public accounts of the Government. They incur expenditure out of these deposits and their receipts are also credited in the same deposits. These deposits of local funds also usually result in a small net balance every year. There are besides quasi or non-governmental associations keeping their transactions as part of government accounts. Lastly come temporary transactions like suspense, remittances etc.

In respect of all these items, the state government acts as a banker and it has the liability to repay. Just as a banker, the state uses all these deposits for its day-to-day ways and means as well as for financing the capital outlay.

Tables 12.1, 12.2 and 12.3 will show the net balance under the heads 'Deposits, Advances and other items' during the period of each Five Year Plan. These deposits are net of transactions on account of purchase and sale of securities but including contingency fund transactions. It will be seen that the state governments have derived substantial benefits from these items for their capital expenditure and for other outlays. It of course also means that the state governments have that much of liability but it is not a liability which will have to be discharged in a particular year or all at once. Hence, even though this liability increases the state governments ignore it for all practical purposes.

It will be seen from the tables that in the First Five Year Plan except for Madhya Pradesh and Uttar Pradesh the other states had overall net balances under this account for the plan period as a whole. In the Second Plan period of all states except Mysore had positive net balances. In the Third Plan period all the states have had positive net balances. Unfunded debts and deposits are thus a source of financing capital outlay and tiding ways and means difficulties even though they are liabilities to the states in the long run. But for these items in the public account, the ways and means of the states as well as their annual overall budgetary position would have less elbow room and a capital outlay of a lesser magnitude alone could have been financed.

We now turn to the role of small savings in the financing of the state Plans. The overall trend in net collection since 1954–55 has shown a more or less progressive improvement except for the short decline registered in the first two years of the Third Plan. The

TABLE 12.1

Net Deposits and Miscellaneous Transactions
First Plan

(Rs. Crores)

Part A States	1951–52	1952–53	1953–54	1954–55	1955–56 (RE)	Total
Andhra Pradesh	—	—	—	0·41	0·48	0·89
Assam	− 2·77	3·38	− 0·45	1·10	0·53	1·79
Bihar	11·83	− 7·16	3·19	1·31	3·37	12·54
Bombay	9·94	− 3·42	1·17	−14·58	6·88	9·49
Madras	20·02	14·20	0·45	0·21	− 5·19	29·69
Madhya Pradesh	− 1·40	− 2·16	− 6·47	− 1·73	− 6·83	−18·59
Orissa	− 0·52	0·39	0·94	− 0·61	2·29	2·49
Punjab	− 0·26	0·35	3·76	0·69	− 1·29	3·25
Uttar Pradesh	− 0·60	14·73	− 3·12	−27·21	−26·48	−42·68
West Bengal	− 0·98	0·41	1·73	2·92	− 1·15	2·93
Part B States						
Hyderabad	6·67	0·39	3·67	− 0·52	− 0·08	10·23
Madhya Bharat	1·18	− 1·72	− 2·37	− 4·52	1·21	− 6·22
Mysore	1·42	− 1·66	− 0·31	− 0·61	− 0·16	1·32
Pepsu	− 0·71	− 0·75	− 0·29	2·30	2·17	2·75
Rajasthan	2·54	− 2·11	1·55	− 1·10	0·50	1·38
Saurashtra	1·54	− 0·23	− 3·14	− 1·03	− 0·43	3·29
Travancore-Cochin	0·32	− 1·26	− 0·46	2·78	− 1·93	− 0·55

[Source: Reserve Bank of India Bulletins.]

mobilization of the resources through small savings depends to a large extent on the efforts made by the State. To give them adequate incentives the proceeds of the small savings collection are shared between the centre and the states. A share of the net collection is given to the states as loan for a period of 10 years carrying interest at 4%. This sharing arrangement was first introduced in 1952, when it was decided that the entire collection of small savings over the targets fixed for individual states should be made over to them in the form of loans subject to the overall condition that the Government of India would retain an amount of Rs. 45 crores out of the Government of India collection for its use. This restriction was however removed in October, 1953, since when states became entitled to receive the excess over the individual target fixed for them irrespective of the total collection realized throughout India. As a further incentive to states for more active collection it was decided during 1954–55 that each state should be given half the excess over 80% of its target, in addition to the half of the collection over its individual target, hitherto received by it. The sharing formula was later on liberalized from time to time until after the declaration of Emergency in 1962 when it was decided that the states' share in 1962–63 and 1963–64 would be subject to a ceiling of Rs. 60 crores and 65 crores respectively. However following the policies of Compulsory Deposit Scheme in respect of land revenue it was decided in November, 1963 that the ceiling would not be applied. Thus, now the collections are divided between states and centre in the ratio of 2:1 without any limitation or ceiling.[1]

TABLE 12.2

Net Deposits and Miscellaneous Transactions
Second Plan

(Rs. Crores)

	1956–57 (BE)	1957–58	1958–59	1959–60	1960–61	Total
Andhra Pradesh	0·12	− 2·46	5·90	6·10	12·01	21·67
Assam	− 0·69	0·86	− 1·22	− 0·20	3·46	2·21
Bihar	10·72	3·04	4·95	8·56	7·42	34·69
Bombay	2·49	− 2·42	− 13·29	—	0·64*	− 12·58
Jammu and Kashmir	NA	0·22	0·29	− 3·18	0·56	− 2·11
Kerala	NA	2·08	0·55	1·26	1·40	5·29
Madras	0·42	− 1·76	6·49	11·79	7·27	24·21
Madhya Pradesh	12·78	3·62	− 0·31	1·33	2·71	20·13
Maharashtra	—	—	—	0·49	0·33	0·82
Mysore	NA	3·15	− 14·66	5·03	4·65	− 1·83
Orissa	0·26	0·08	− 0·92	4·81	4·09	8·32
Punjab	4·19	2·12	2·86	− 4·55	1·73	6·35
Rajasthan	NA	0·36	2·39	6·29	2·97	12·01
Uttar Pradesh	− 18·60	3·51	− 13·21	− 5·22	− 4·03	− 37·55
West Bengal	0·16	7·32	− 11·49	2·56	3·57	2·12
TOTAL	11·85	19·72	− 31·67	35·07	48·78	83·75

*Gujarat.

[Source: Reserve Bank of India Bulletins.]

[1]Cf. Reserve Bank of India Bulletin, September, 1966. See also, V. G. Mutalik Desai, Savings in a Welfare State (Manaktalas, 1966.)

TABLE 12.3

Net Deposits and Miscellaneous Transactions
Third Plan

(Rs. Crores)

	1961–62	1962–63	1963–64	1964–65 (RE)	1965–66 (BE)	Total
Andhra Pradesh	5·28	12·16	9·34	10·31	8·60	45·69
Assam	0·72	– 0·81	1·90	– 0·02	– 0·39	1·40
Bihar	6·30	1·27	– 5·21	– 1·51	0·94	1·79
Gujarat	5·03	4·63	– 1·49	4·07	4·73	16·97
Jammu and Kashmir	1·22	1·14	– 0·30	0·08	5·01	7·15
Kerala	– 0·31	2·11	2·02	0·12	1·40	5·34
Madras	6·19	1·93	11·44	5·79	6·91	32·26
Madhya Pradesh	0·51	5·47	1·76	3·86	3·60	15·20
Maharashtra	10·22	15·08	12·67	15·89	7·02	60·88
Mysore	6·15	5·18	4·08	4·37	5·30	25·08
Orissa	4·04	10·12	– 4·26	12·44	6·17	28·51
Punjab	2·05	5·10	1·72	12·61	10·52	32·00
Rajasthan	2·94	3·22	7·81	– 1·91	7·50	19·56
Uttar Pradesh	7·29	3·72	6·97	28·57	29·23	75·78
West Bengal	5·18	7·93	5·69	5·01	6·41	30·22
TOTAL	62·81	78·25	54·14	99·68	102·95	397·83

[Source: Reserve Bank of India Bulletins.]

There are three main sources of small savings viz. the Post Office Savings Bank, the National Defence or Savings Certificates and the Cumulative Time Deposit Scheme. The Post Office Savings Bank is quite suitable for rural areas and for small amounts but the collections can be neither regular nor certain. The Cumulative Time Deposit scheme is particularly suitable for the salaried class who can invest a small sum regularly every month. The certificates are suitable for *ad hoc* or institutional investors and provident funds.

Table 12.4 will show the small savings collection during the Third Plan period by the states.

Small Savings collections are concentrated in the six states of Maharashtra, West Bengal, Uttar Pradesh, Madras, Bihar and Gujarat which among themselves accounted for 65% of the net collection in 1965–66. Among the states, Maharashtra tops the list in the matter of total collection in all the five years, with a share ranging from 15% in 1965–66 to 27% in 1962–63. Except for 1961–62 in all the other four years West Bengal occupies the second place followed by Uttar Pradesh. Post Office Savings Banks collections are comparatively more prominent in agricultural states like Uttar Pradesh, Punjab and Bihar whereas National Savings Certificates are favoured more in industrially advanced states like Maharashtra, West Bengal and Gujarat.

The increase in small savings collection over the Second Plan period was contributed mainly by Maharashtra, West Bengal and Gujarat, followed by Uttar Pradesh, Madras, Bihar, Mysore and Punjab. Andhra Pradesh and Rajasthan showed a deterioration in collection when compared to the Second Plan.

It is quite clear that if states would adopt right methods to increase small savings collections in their areas there will be considerable resources mobilization for the Plan. The Third Plan period has shown rather promising results which will require to be

TABLE 12.4

Small Savings Collections by States in Third Plan

(Net, Rs. Crores)

	1961–62	1962–63	1963–64	1964–65	1965–66 (provl.)
Andhra Pradesh	2·70	−0·15	2·16	1·94	3·79
Assam	4·30	3·99	4·49	4·68	4·36
Bihar	7·57	8·04	9·89	11·38	11·88
Gujarat	9·08	2·80	9·24	10·35	7·63
Jammu and Kashmir	0·85	0·39	0·66	1·04	1·14
Kerala	1·68	0·97	2·47	2·00	3·98
Madhya Pradesh	2·72	2·50	5·56	4·64	5·55
Madras	2·34	2·09	5·91	6·29	14·70
Maharashtra	18·52	20·60	29·55	26·20	22·12
Mysore	2·51	2·24	4·55	4·35	5·88
Orissa	2·08	2·73	2·94	3·22	2·85
Punjab	4·68	4·58	8·15	8·85	13·56
Rajasthan	1·12	0·38	1·51	1·52	3·21
Uttar Pradesh	12·78	7·61	14·06	17·10	18·42
West Bengal	11·67	13·69	22·16	21·34	21·10
TOTAL	84·50	72·47	123·28	124·91	139·16

Note: 1. The total will not tally because of rounding off under states.

2. The figures by themselves do not convey much about the performances or potentialities of states unless related to such factors as state income, population and urbanization.

[Source: Reserve Bank of India Bulletin, September 1966, pp. 1044–5.]

followed up intensively in the Fourth Plan period. The Small Savings movement should however spread to all states instead of being concentrated in particular states. The large collections in Small Savings in West Bengal and Maharashtra seem to be largely helped by the investments of provident funds of companies. While this no doubt contributes to an increase in small savings figures, it does not result in a net addition to the savings of the community. The hard and urgent task is to inculcate a real feeling of thrift in all sections of the population and make the savings effort a steady and systematic one. There have been considerable difficulties in this regard not least of them being the rather unsatisfactory rate of interest and comparatively cumbersome procedures. Considerable attention is being paid now to spreading the small savings movement and it will not be too much hope that better results will ensue in the coming years.

We now turn to a study of the financial aspects of the State Electricity Boards. Power generation has, as we have seen already, assumed a major role in the developmental efforts of the state governments and provides the key to agricultural as well as industrial development. Electricity is a subject in the Concurrent List. The State Electricity Boards are constituted under a Central Act, namely, 'The Electricity (Supply) Act 1948'. These Boards operate under the overall policy control on the state governments which have, it is needless to emphasize, a vital interest in the sound working of these Boards. At present the growth of power development and of the electricity boards themselves varies from State to State.[1] For example in Assam, Bihar and certain other states, development of power generation under the auspices of the electricity boards is just under way while in states like Madras, Madhya Pradesh, Punjab and certain other states, the electricity boards already manage a large complex of generating stations and transmission and distribution systems. Although Maharashtra and West Bengal are well advanced industrially and the growth of power development has generally kept pace with the needs of industrial expansion in those states, the private licensees are in command of most of the developed and paying areas while Gujarat to a large extent depends on thermal power supply, resulting in fairly stiff rates. Punjab, Kerala and Mysore are in a more favourable position as cheap hydro supply is available in these states. Madras, Punjab and Andhra Pradesh are well ahead in the rural electrification programme. While some state governments have retained the construc-

[1]Report of the Committee on the working of the State Electricity Boards, 1964, p. 3.

tion of power projects in their hands, others have entrusted the responsibility to the state electricity boards.

The capital structure of the state electricity boards is made up from three sources at present, namely, loans received from the state governments, the open market borrowings and the internal resources of the Board (i.e. accumulations in the reserve and depreciation funds). The loans from state governments however predominate in the capital structure. While advancing the loans under Section 64 of the Electricity Supply Act 1948 the state governments prescribe the terms and conditions on which the loans are sanctioned by them from time to time. Some of the state governments have prescribed the terms of repayment of the loan in addition to the rate of interest etc. while others have not made any stipulations regarding the repayment of their loans. In fact Section 67 of the Act which deals with the allocation of the revenues of the boards does not provide for repayment of the loans received from the state governments from the revenues of the boards. While Section 68 permits to a certain extent the repayment of the open market loan floated by the boards from out of the accumulations in the depreciation reserve fund, the said fund cannot be drawn upon to repay the loans from the state government. Thus, under the Act as it stands at present, the boards could repay the loans to the state government either by floating open market loans or by getting fresh loans from the state governments themselves. Otherwise these loans will have to be treated as perpetual loans by the state governments without expecting any repayment from the boards. The state governments however will have an obligation to repay the loans received from the central government as also their open market borrowings.

The Act does not provide for repayment of loans from revenues presumably because the depreciation reserve created could be deemed to serve the purpose of replacing the assets when they are worn out. The cost on assets has however progressively been on the increase and the depreciation fund provided as per the provisions of the Act has proved to be inadequate to meet the replacement cost.

This description of financial structure of the electricity boards will clearly convey the extent of impact of these boards on state finances and the extent to which this impact is likely to increase in future in many states. As a matter of fact some boards have found themselves unable even to pay interest charges on the loans advanced by the state governments. In fact the only boards which were in a positon to pay these charges in 1962–63 were those of Madras, Mysore, Madhya Pradesh and Uttar Pradesh. Table

12.5 will show the capital structure of the electricity boards and the percentage of return on loan capital. It will be seen that the financial position of many of the boards is quite unsatisfactory.

It is clear that the state finances are vitally affected by the working of the electricity boards to the extent that the boards are not able to pay the interest charges on loans. The state governments have to pay the interest charges from their own general revenues. Besides any modification of the electricity tariff has a double-sided concern for the state government. For one thing, such an increase would be welcomed by the state governments to the extent that it

TABLE 12.5

Financial Returns of Electricity Boards (1962–63)

(Rs. Crores)

Name of Electricity Board	Capital assets	Loans from Government	Loans from Public	Internal Resources	Percentage of returns on Loan Capital*	Percentage of returns on Government Loan†
Andhra Pradesh	83	68	8	2	1·10	0·49
Assam	17	16	1	—	−0·98	−1·28
Bihar	52	46	—	6	1·50	1·50
Gujarat	58	47	8	3	1·69	1·50
Kerala	59	44	4	11	2·84	2·71
Madras	170	129	11	30	5·67	5·82
Madhya Pradesh	55	53	—	—	4·90	4·90
Maharashtra	54	39	6	9	1·05	0·74
Orissa	20	18	2	—	0·46	0·28
Punjab	116	105	—	11	4·70	4·70
Rajasthan	32	32	—	—	0·60	0·60
Uttar Pradesh	96	96	—	—	4·20	4·02
West Bengal	23	21	4	1	−0·78	−1·27

Note: In some cases total of resources does not tally with capital assets.

*Before payment of interest and excluding government subsidies.
†After payment of interest on the public loan and including government subsidies.

[Source: Report of the Committee on the working of the State Electricity Boards, Government of India, 1964, pp. 16–17.]

would enable the boards to pay their interest charges and thus reduce the unnecessary burden on general revenues. At the same time to the extent that the tariff is increased, the scope for increasing the tax on consumption of electricity or other duties on electricity is curtailed. State governments will therefore have to choose between alternative methods of effecting an increase in the price of electricity. Such of those states which are not able to get back the interest charges from the electricity boards will no doubt prefer an increase in the tariffs to an increase in the tax while those who get the interest charges would prefer to increase the tax rather than the tariff. The rates of electricity tariffs themselves vary from state to state and from use to use. Separate rates are fixed for high tension and low tension supply and for different uses like industrial, commercial, domestic and agricultural. State governments have also come forward to supply electricity at concessional rates as an incentive to new industries. A rationalization of the rate structure is also quite necessary.

In view of the different stages of power development in various states, the question will arise whether it would not be desirable to have a national set up for power generation instead of leaving it to individual states. The advantages would of course be uniform rates, better co-ordination and choice of power programmes with the least cost and also a national grid but states are likely to contend that such a set up may not redound to maximum regional development. The disparities in the level of power generation in various states would also produce conflicts of interest. While therefore a national setup for power generation is desirable in the financial sense it may not find ready acceptance among states. In any case the disparate levels of development and financial structure of electricity boards do seem to require thinking on national lines.

The first step would of course be to make them pay their interest charges. A Committee of Ministers which went into the working of the State Electricity Boards in 1964 considered that the minimum object should be to earn revenues sufficient to cover operation and maintenance charges, contribution to depreciation and general reserves and interest charges on loans. The quantum of contribution to the reserve would also require enhancement. The objective of covering the charges mentioned above should be achieved within a period of 3 to 4 years. As a second phase objective a balance of revenue working out to a net return of 8% on the capital base should be achieved within a period of 3 to 5 years after the achievement of the first phase. The rate of interest charged by the state governments on their loans should be fixed following a uniform principle.[1]

A major question is of course why the loans from the state governments to the boards should be perpetual since this adds to the debt burden of the states but the Committees decided that no change in the existing capital structure to the state was necessary. Actually, in the context of the inevitability of non-productive outlay in certain other sectors, there is no reason why the electricity boards should not be insisted upon to pay their way. But the Committee should have thought that there was no point in thinking of the principal when the interest itself was in danger of default!

We shall now devote ourselves briefly to the problems of financial administration and budgetary reform relevant to the finances of state governments. These are problems for which attractive solutions can be put forward. But it is not necessary to do so since radical reforms are not likely to be accepted overnight in these fields. What is more, the state governments in India do have a well-established and fairly satisfactory system of financial administration and accounting and it is not as if the states' finances and their management have unduly suffered for the lack of a proper system of administration and budgeting. Within the dimensions of this work we shall not be able to make any observations more than in the nature of *obiter dicta.*

The financial administration of a state government is comparatively simple. There are no decisions of economic policy involved as in the centre. By and large the main concern is to keep a reasonable balance between receipts and expenditure. But, with the advent of planning, it has become highly necessary that the finance departments of state governments are imbued with a sense of responsiveness to the needs of development. In other words, the art of finance is not simply one of husbanding resources but of spending them as well. Such an attitude cannot of course arise *ex nihilo* and it will take some time before financial administration in general comes to grips with the needs of planning. An old-fashioned Finance Department will easily raise objections to proposals of the kind it is accustomed to rejecting, like the expenditure on the appointment of clerks and so on. But it may simply yield, in truly Parkinsonian fashion, to big and high sounding proposals which it may not fully understand. So far as the needs of planning are concerned, the attitudes of the Finance Department will have to pass through various

[1]Pp. 7–10 ibid.

stages of evolution. Probably the first step will be for the Department to be responsive to planning needs and sanction plan schemes without delay. But outright acceptance can be as bad as outright rejection because a scheme cannot be accepted per se merely because it bears a plan label. An evolved Finance Department would therefore have the combination of a sense of perspective and responsiveness to planning needs with the need for economy and the dictates of common sense. It is not possible to theorize on the subject, for a lot will depend on the kind of persons who man the department and the experience they have had in various fields.

So much depends in administration on the approach of the persons concerned that the question of merging the Planning and Finance Departments is a question that cannot be answered in absolute terms. Some states have a combined Finance and Planning department and others have separate departments. Each state claims that its system is working smoothly. But there seems to be a clear advantage in having the Planning and Finance Departments together in that it rules out the consequences of a Planning Department being frustrated by an unresponsive Finance Department.

Evaluation has never been the strong point of many a financial administration. There is an urgent need for financial administration to actively involve itself with the evaluation of schemes and projects. This in turn would mean the availability of an adequate statistical base which cannot be said to exist in many states. The concept of the functions of the Public Accounts Committee will also have to change in the sense that it should concern itself less with mere questions of excess or short falls of expenditure over appropriation and more with the results of various schemes. No doubt the Estimates Committee is also there but the Public Accounts Committee itself can adopt a more comprehensive approach to the scrutiny of Public Accounts. The grants-in-aid to the local bodies have also become so considerable in magnitude that a concern over their proper utilization is legitimate. This would mean that the Public Accounts Committee and the Legislature itself should be supplied with sufficient information to assess the results of government grants to local bodies.

The emphasis on evaluation inevitably implies a certain change in approach and addition to the work of the Comptroller and Auditor-General and his Accountants-General, who keep the accounts of state governments and audit them as well. In recent years the Accountants-General have detailed some staff in their organizations for 'efficiency-cum-performance audit'. In Uttar

Pradesh a bigger organization has been established by the Planning Commission as an experimental measure for evaluating the results of plan schemes.

When problems of budgetary reforms are raised there is a temptation to get bogged in concepts like programme budgeting and performance budgeting. As already stated, the state governments have a reasonable accounting and budgeting base in position. The first step is to streamline the existing organization rather than try to devise an altogether new budgetary format or procedure. The management and administration division in the Committee on Plan projects of the Planning Commission is reported to be examining the question of linking of heads of development and budgetary and accounts heads and also the question of the introduction of performance budgeting in India.[1]

It is necessary first of all to have better reporting and better presentation of the existing material. We have already seen how the classification of plan schemes is not and cannot be on all fours with the budgetary classification of account heads. The first need is to publish an intelligible co-relation between plan schemes and budgetary heads—which can be done by a separate publication as in Madras State and presented to the Legislature along with the budget. It is also easy to introduce elements of programme and performance budgeting in the budget memorandum or by means of a separate publication. This can be done even now without waiting for some spectacular budgetary reform. Again, with the increasing financial aid to local bodies, a separate brochure on grants-in-aid to them will be necessary. These are now scattered under various budget heads and cannot be readily picked out by laymen. Such a brochure can also report on the state of local body resources for the information of the Legislature.

These are additions to budget documents which will present the budgetary figures in a more intelligible form to legislators, lay men and administrators themselves. At the same time, it is possible to reduce the forbidding bulk of the budget documents. The detailed budget estimates can be made to look less frightening by some simple reforms. The simplest seems to be the abolition of certain unwanted detailed heads. According to the accounting classification that is in force, the expenditure on staff is booked under several detailed heads like pay of officers, pay of establishment, allowances (Dearness Allowance, Travelling Allowance and other Compen-

[1]See ECAFE *Bulletin*, Sept., 1966.

satory Allowances) and office contingencies, Dearness Allowance having very much come to stay, it is not clear why it should be accounted separately. It cannot be said that such detailed figures are necessary either for reporting or for control. They are not necessary for reporting because the same figures can be obtained with reference to the number of statements giving the number of staff of various categories employed in various departments. With reference to these, estimates like the estimated cost of additional dearness allowance can be computed. These figures are also not necessary for purposes of control. When differing scales of pay, dearness allowance etc., are clearly laid down, it is not possible for any head of department to spend more on pay and allowances. Detailed figures relating to this can be asked to be prepared by each head of department but they need not necessarily be presented in the budget estimates. Hence items like pay of officers, pay of establishment, Dearness Allowance and other Compensatory Allowance, pay of contingent staff and Dearness Allowance to contingent staff, special allowance to contingent staff can be merged in one head namely 'Salaries and wages'. This simple abolition of a few detailed heads will cut down the size of the detailed budget estimates by about 25 to 33%. There is need for control on expenditure on items like travelling allowances and office contingencies which can continue to be exhibited separately.

The economic classification of the state budgets has become necessary since it is essential for an understanding of the impact of government outlays on the economy. Besides the central government has, for nearly the past ten years, been preparing an annual booklet on such an economic classification, presented along with budget documents. Some states like Madras have attempted economic classification of their budgets from time to time. The Punjab University and the National Council of Applied Economic Research have also attempted economic classification of state budgets. It is necessary to have a common and accepted form of economic classification both at the centre and the states so that the figures can be compiled for the whole of India without any difficulties in compilation.

What we have said about budgetary reforms will look fairly obvious and simple but many state governments cannot be said to have exercised their minds on these questions. There seems to be, besides, scope for improvement in the exchange of information among states. We have suggested elsewhere institutions for inter-governmental co-operation and co-ordination. Along with such institutions, and even separately, there is a great need for a small body to study and compare the finances of state governments in sufficient detail. It is not necessary to have any highsounding institution. Even if each state agrees to contribute say Rs. 500 per month as a grant to a Research Body like the National Council of Applied Economic Research it will be possible to have a Director of Study and a few research officers to compile, compare and study the finances of state governments to an extent that will surely be beneficial to all the States.

CHAPTER XIII

STATE LOCAL RELATIONS

If state governments complain that they are the Cinderella of national finances, the local governments can complain that they are the Cinderella of Cinderellas. Only there is no handsome prince in sight! The importance of local finance is apt to be neglected, particularly in a federation. With the establishment of Panchayat Raj institutions in India, the problem has assumed a certain urgency and importance and it is necessary to deal with it in some detail first. We shall deal with the problems of urban local bodies later in the chapter.

What are the reasons for the establishment of local governments?[1] If they require the achievement of specific aims, should they not have the requisite financial endowment for this purpose? To begin with, a reason that is often important is that local governments serve as training grounds in democracy, as the primary schools of politics. It is easy to panegyrize this or shrug it off as naive self-deception. But it ought to be agreed generally that local bodies could serve as springs of leadership and also enable national leadership to be broadbased. Could this aim be achieved, if local bodies remain only in name, with no financial endowment to provide them with substance? Lack of finance and consequent atrophy will only make such leadership as emanates feel a sense of impotence. It will make it think that the art of local government does not lie locally but consists in pulling strings, at the state headquarters for increased financial allotments. Frustrated leadership finds it easy to flow into short cuts and devious channels; it gets an early taste of corruption which it carries as it goes up the political ladder. When finances are limited, provision of services becomes a matter of communalism or politics. Counter-pressures develop and local affairs begin to assume a disproportionate controversial importance. People in their turn tend to look upon local government as a device to feather someone's nest and their political horizon being limited, they tend

[1]See the author's article 'Local Finance in Developing Countries' in *Journal of Local Administration Overseas*, London, July, 1965. The next few pages substantially follow the argument in that article.

to view the national government in a similar way. One is not saying all this happens, or will happen everywhere. But if we neglect local finance largely, when at the same time we choose to have local governments, we are leaving the door wide open for such possibilities.

Or take the other aspect which receives wider acceptance as a justification for local government—the provision of services. It is admitted on all sides that a central bureaucracy cannot provide essentially local services to the satisfaction of the people (which is more true in developing countries with a dearth of administrative talent); the discretion in detail has to be a local discretion. In the context of the 'revolution of rising expectations' the provision of such services assumes vital importance; in the short run national development plans, to the rural population, will be synonymous with the provision of these services. There is a great demand in developing countries for the provision of health facilities and primary education, and no realistic planner can afford to ignore these social overheads. Most of these services have to go to local government hands sooner or later, but at the same time they involve large outlays, often on a scale the local bodies have never handled before. Here we come across the double-edged problem of local finance. On the one hand, there has to be provision of sufficient aid to local bodies for the satisfactory maintenance of the services and the attainment of plan targets; on the other, there have to be adequate safeguards to ensure that the money is spent properly and the best results are achieved.

There is one connected aspect in which local finance has to keep in step with development. That is the problem or urbanization—rural areas becoming urban and small towns becoming large towns. Urbanization brings problems as well as benefits and these problems, such as sanitation and housing, have to be dealt with concurrently as they arise, if not planned ahead. These are essentially local fields. In such cases the financial base of the local bodies has to be expanded and more grants-in-aid will have to flow. Very often, rural and urban local bodies have different financial set-ups and regulations governing grants-in-aid. The urbanizing local bodies would therefore have to be properly 'classified' and taken care of if the problems of urbanization are to be kept in check; and they may require a 'most-favoured' financial deal from the central government. In this sense, local finance cannot wait for 'better times'.

We now come to the third important aspect from which local

bodies can be looked at from the national level. That is, local bodies can in some respect be engines of economic development. Though it is true in a limited sphere only, it is better not to underestimate its importance. For one thing, economic development plans for rural areas often consist of a series of small projects spread throughout the country, and only local bodies can be agencies for implementation of such projects. The building of village roads, the provision of wells and the construction of primary school buildings are all cases in point. Or take the case of agriculture, where many improvements which can substantially increase productivity require only little application of capital, but have to be brought home to the farmer. Such extension work is likely to achieve better results under the aegis of local bodies, for the influence and example of prominent local people will be behind it.

In a wider and more general sense, local bodies are valuable instruments for the provision of 'external economies' to the rural peasant economies. They thus help accelerate the transition from the subsistence economy to an exchange economy. The building of village roads and the consequent enlargement of the market is a case in point. Local bodies can help mobilize capital formation also. We have the suggestion of the late Ragnar Nurkse that disguised unemployment has disguised savings potential; that if the surplus farm workers are siphoned off and can be made to work on the building of roads, etc., there would be a new source of capital formation. This suggestion may no doubt have practical difficulties, but if there is any institutional arrangement at all that can secure such mobilization, it is the local bodies.

Thus there are effective ways in which local bodies can be used for the cause of economic development. There is this belief, implicit and unexpressed, in the many-sided activities of the Panchayati Raj institutions in India. But it is as well to remember that we are only at the periphery of this aspect of local government. We do not have clear-cut examples of the process of economic development being substantially accelerated through local governments. Nor are we sure in any way of the manner in which the spark of enthusiasm and activity could jump from the planner to the local elite and from the local elite to the people at large. But there is no doubt that there is a vast reserve of local enthusiasm that can be harnessed for economic development and the problem is to find the key to the flood-gates. The strategy of development, as Hirschman has impressed on us, consists in setting up pacing devices, pressures and inducement mechanisms that can elicit hidden reponses. A search for such devices *vis-à-vis* local bodies, and a study of the motivation between local bodies and the local population will definitely be rewarding.

Thus there is a sense in which the money spent on local bodies can be considered to be an enormously productive investment rather than as an unavoidable item of expenditure of a low priority, to be grudgingly allowed with a cynical shrug of shoulders. Local bodies can in fact be viewed as 'multi-purpose economic institutions'. From this angle, local finance will be seen in an entirely different light.

It may be argued that the claims made for local bodies are based on what they might be, rather than on what they are. The answer is that they are what they are partly because they never receive a fair financial deal. Besides, there is no straight and narrow path to economic development and it can never proceed without experimentation; no one is sure how rural peasant economies could be activated and local bodies could be a possible *primum mobile*.

We may now turn to an examination of the position of local bodies in India with particular reference to their relations with the state governments and the manner in which the state governments can help and guide their activities. We shall first deal with the rural local bodies, though much of what we will say will also be relevant to urban local bodies.

Schemes of rural development were initiated in all the states in the First Five Year Plan period with a programme known as the programme of Community Development and National Extension Service which was substantially aided by the centre. It was soon recognized that such a programme of rural development cannot thrive except in a climate of 'democratic decentralization'. The term Panchayati Raj refers to the pattern of rural local government established in recent years in India in view of the felt need for democratic decentralization and in pursuance of the Directive Principles of State Policy embodied in the Constitution, which provide, *inter alia*, for the establishment of panchayats all over the country. Most state governments have enacted panchayati raj legislation but the pattern of local government varies from state to state. Some states have three tiers of local government and others two. The nomenclature of the local body also varies in certain cases. For the purpose of this work we could stop with saying that the term Panchayat refers to the local government institution at the village level, the term Panchayat Samithi refers to the local government institution for an area of roughly 80,000 population and covering usually a number of village panchayats. The term Zilla Parishad refers to the local government institution at the district level.[1]

There are considerable variations among the states in the matter of allocation of functions and finances to these local bodies, but the following may be said as a broad description of the position. Panchayats usually have the functions of formation of village roads and their maintenance, village sanitation, conservancy and water supply, lighting and certain optional functions as also schemes which state government may require to be implemented through the panchayats. The panchayat samithi is in many states co-extensive with the 'development block', that is an area of roughly 80,000 population (a concept introduced by the Community Development programme) and it has the functions of promoting elementary education, maintenance of dispensaries and health centres, formation and maintenance of inter-village roads, schemes under the Community Development programme and other activities of a local or developmental nature. The zilla parishads have not been established in all the states and where they are, they sometimes perform some of the functions of samithis like elementary education, besides the maintenance of roads at district level, health programmes, secondary education and so on. A zilla parishad in Maharashtra is a miniature state with a wide repertoire of functions.

As regards finances, the local bodies depend very substantially on government grants and the funds under the Community Development programme which are given by the centre. This apart, the income of a panchayat is usually derived from property taxes and from a statutory share in land revenue. In most states the panchayats also levy profession taxes and vehicle taxes and in certain states like Madras they get a share of entertainment tax and stamp duty on transactions on immovable property. In order to ensure that panchayats really exercise their powers of taxation, levy of one or more of the local taxes namely house tax, profession tax and vehicle tax has been made compulsory by legislation in some states. The panchayat samithis get their funds from the Community Development programme and besides have often the power of levy of taxes particularly on land revenue. They also get a share of land

[1]As on March, 1964, there were over 200,000 panchayats in the country covering 533,000 villages and a population of 283 millions (which is about 95% of the rural population). The number of samithis will eventually be around 5,000. See pp. 20 and 23 of *Community Development*, Publications Division, Government of India, 1964.

revenue and/or the cesses on land revenue. The financial base of the zilla parishad is made up largely of shares of land revenue and grants-in-aid.

Taking the local bodies as a whole their financial resources can be classified into following groups: (1) purely local taxes like house tax, profession tax and vehicle tax which are usually levied by the panchayats; (2) independent power of taxation by way of surcharge on land revenue or entertainment tax; (3) taxes levied and collected by the states but distributed to the local bodies like entertainment taxes and surcharge on stamp duties; (4) devolution of land revenue and/or cesses on land revenue and; (5) grants-in-aid of various kinds. With increasing accent on developmental schemes the expenditure of local bodies has grown substantially. The quantum of government grants to the panchayati raj institutions is now a substantial portion of the states' revenue expenditure as a whole and of their development expenditure in particular, as will be seen from Table 13.1.

Table 13.2 shows the position in 1963–64 in respect of the three states where the quantum of government grants was the highest.

Though the local bodies as a whole are now receiving much more resources than they ever did, the individual financial position of

TABLE 13.1

State Grants to Local Bodies

(Rs. Crores)

Year	Quantum of Grants to P.R. Institutions	Total expre. of States	Percentage of (2) to (3)	Total dev. expre. of States	Percentage of (2) to (5)
(1)	(2)	(3)	(4)	(5)	(6)
1961–62	77·05	1,042·00	7%	615·16	13%
1962–63	133·99	1,170·97	11%	673·06	20%
1963–64	138·94	1,269·16	11%	727·54	19%

[Source: Column (2) from Report of the Study Team on Panchayati Raj Finances, p. 73. Columns (3) and (15) from Reserve Bank of India Bulletins. In figures in column (2), those for 1961–62 and 1962–63 are actuals and those for 1963–64 are budget estimates.]

Note: Jammu and Kashmir and Gujarat have been excluded as figures under column (2) are not available.

TABLE 13.2

State Grants to Local Bodies in Some States
1963–64

(Rs. Crores)

Name of State	Grants to P.R. bodies	Total expre. of State	Percentage of (2) to (3)	Total dev. expre. of State	Percentage of (2) to (5)
(1)	(2)	(3)	(4)	(5)	(6)
Andhra Pradesh	26·73	113·91	23%	69·89	38%
Madras	29·42	120·77	24%	77·99	38%
Maharashtra	41·15	156·88	26%	71·56	58%

[Source: As for Table 13·1].

many of the local bodies cannot be said to be satisfactory. Particular care has not been taken to match the functions to be discharged by the local bodies with the finances needed for the purpose. All the local bodies have been created by state statutes. But such legislations have not taken into account the need for giving the local bodies the sound financial base which alone can support the load of responsibilities cast on them. The Study Team on Panchayati Raj Finance, pointed out in 1963: 'We cannot help wishing that when the Acts were passed by the legislatures, an attempt had been made to estimate the minimum cost of fulfilling the obligatory functions and to provide resources for the purpose. The actual resources often vary inversely to the number and extent of obligatory functions'.[1] Table 13.3 will illustrate the levels of *per capita* income of the various local bodies and their *inter se* variations within and among states.

It would be wrong to conclude however that the financial plight of local bodies has gone unnoticed. India has a long tradition of rural and urban local government though it is only in recent years that the entire country has been covered by panchayats (or municipalities). The Local Finance Enquiry Committee, 1951, went into the financial problem of local bodies in great detail. It recommended

[1]Page 10, Report of the Study Team on Panchayati Raj Finances (1963), Part 1.

thirteen items of revenue[1] (including one in the Union List, *viz.*, terminal taxes on goods or passengers carried by railway, sea or air to be exclusively set apart, by convention, for exploitation by local authorities. Local bodies should have a free hand, within prescribed maxima, for determining the rates of taxes. Where a local body was unwilling to impose a tax at an adequate rate the state government should have the right, in the first instance, to give friendly advice and if the local bodies fail to carry it out the state government should in the last resort have the power to impose or raise the tax themselves. The Committee preferred assignment of revenues to grants-in-aid which should however come in to remedy inequalities inherent in a system of assigned revenues.

[1] UNION LIST

(1) Item No. 89.
Terminal taxes on goods or passengers carried by railway, sea or air.

STATE LIST

(2) Item No. 49.
Taxes on lands and buildings.
(3) Item No. 50.
Taxes on mineral rights subject to any limitations imposed by Parliament by law relating to mineral development.
(4) Item No. 52.
Taxes on the entry of goods into a local area for consumption, use or sale therein.
(5) Item No. 53.
Taxes on the consumption or sale of electricity.
(6) Item No. 55.
Taxes on advertisements other than advertisements published in the newspapers.
(7) Item No. 56.
Taxes on goods and passengers carried by road or on inland waterways.
(8) Item No. 57.
Taxes on vehicles, other than those mechanically propelled.
(9) Item No. 58.
Taxes on animals and boats.
(10) Item No. 59.
Tolls.
(11) Item No. 60.
Taxes on professions, trades, callings and employments.
(12) Item No. 61.
Capitation taxes.
(13) Item No. 62.
Taxes on entertainments including amusements.

The Taxation Enquiry Commission, 1953–54 also considered local taxes in considerable detail. It recommended six items of revenue[1] for exclusive utilization by or for local bodies, besides four others for permissive utilization by them. It also recommended that not less than 15% of land revenue and 25% of motor vehicles taxes should be distributed to local bodies. However it preferred grants-in-aid to assignments of shares of taxes as a method of financing local bodies. It recommended the adoption by each state government of a system of grants-in-aid based on the following principles:[2]

TABLE 13.3

Variations in levels of income of local bodies
(1961–62) (Rupees per capita)

State	Panchayats		Panchayat Samithis		Zilla Parishads	
	Highest	Lowest	Highest	Lowest	Highest	Lowest
1. Andhra Pradesh	8·04	0·74	9·37	2·99	6·09	3·92
2. Bihar	0·60	0·08	—	—	—	—
3. Madras	1·02	0·48	9·86	5·36	—	—
4. Maharashtra	5·41	2·00	4·86	1·27	18·37	8·64*
5. Mysore	3·73	0·72	1·32	0·38†	—	—
6. Orissa	4·09	1·00	8·53	5·10	—	—
7. Punjab	9·20	1·15	—	—	—	—
8. Rajasthan	7·15	0·79	6·00	4·10	0·04	0·02
9. Uttar Pradesh	3·74	0·10	2·19	0·81‡	2·02	1·34
10. West Bengal	—	—	2·03	1·05§	—	—

[Source: The figures have been compiled from Part II of the Report of the Study Team on Panchayati Raj Finances, 1963, Ministry of Community Development and Cooperation. The study team wanted to visit institutions with generally the highest and lowest incomes. On that basis, the figures are furnished in the Report as highest and lowest.]

*Figures for 62–63. †Taluk Development Boards.

‡Per Capita Expenditure. §Anchal panchayats.

[1](1) Taxes on lands and buildings;
(2) Taxes on the entry of goods into the area of a local authority for consumption, use or sale therein, popularly known as octroi;
(3) Taxes on vehicles other than those mechanically propelled;
(4) Taxes on animals and boats;
(5) Taxes on professions, trades, callings and employment; and
(6) Taxes on advertisements other than advertisements published in the newspapers.

[2]p. 366, Report of the Taxation Enquiry Commission, 1953–54, Vol. III.

(1) There should be a basic 'general purposes' grant for each local body other than the bigger municipalities and corporations;

(2) the local bodies eligible for such grant within each category (municipality, local board, panchayat, etc.) should be classified into a few simple divisions based on population, area, resources etc. and the grant itself related to these factors as well as to the size of the normal budget of the local bodies;

(3) the basic grant should be such that, after taking into account its own resources, the local body will have fairly adequate finance for discharging its obligatory and executive functions;

(4) the basic grant should be assured over a reasonable number of years—say three or five and, save for exceptional reasons not be subject to alterations from year to year within that period; and

(5) there should in addition be specific grants (annual and other) which, as at present, will be for particular items and services. These should be conditional on (a) the particular service being maintained at a prescribed level of efficiency and (b) the local body exploiting its own resources to the extent indicated by government from time to time.

Before we proceed to briefly narrate the recommendations of the Study Team on Panchayat Raj Finances we may give a broad description of the financial position of local bodies at the three tiers of panchayati raj. Statistics are not available to give a reasonable statistical picture, though we will attempt to illustrate the position here and there.

The largest disparity between functions and finances is in the case of panchayats. The financial position of a large number of them is very poor and may not exceed one Rupee per capita. At the same time their functions are many and their tasks formidable. The provision on drinking water supply, drainage and sanitation and the formation of village roads are basic to any programme of rural development and have to be subsidized to a large extent by the state and central governments. Grants for these purposes take the form of matching grants, to be matched by contributions from the local public or the panchayat. Needless to say, the requirements of matching contribution puts financially better placed panchayats to advantage in securing these amenities in a larger measure.

The panchayats are also handicapped by the necessity to pay salary to a part-time or full-time Secretary—which sometimes absorbs more or less the entire revenue and sometimes by their uneconomic size from the point of view of financial viability.

As regards the panchayat samithis, the gap between the needs and resources is not as much as in the case of panchayats but it does exist. Substantial differences also exist between the resources of one samithi and another. Panchayat samithis (unions) in Madras State have a creditable record of taxation. But the samithis have a vast field of responsibilities before them. Elementary education is an onerous duty involving substantial and growing expenditure. The samithis are also pivotal points in the implementation of the Community Development programmes. The maintenance of roads and dispensaries too are items where they have to incur substantial expenditure. The samithi is an important level of local administration and the availability of finances or otherwise can make or mar the development of the area. The need for a basic minimum of financial opportunity at the samithi level deserves to be recognized in all the states. The problem of disparities in samithi resources has also arisen wherever the resources are land revenue based since the quantum of land revenue will vary from area to area depending on the fertility of the soil and other factors.

As regards zilla parishads the scope varies from state to state. At the one end is the Maharashtra State where the parishads spend about one-third of the total revenue budget of the state and at the other are district bodies with no executive functions and in between are bodies whose budgets are on different level of expenditure. The Study Team on Panchayati Raj Finances came to the correct conclusion when it said: 'In the face of such wide variations we do not think it is necessary or useful for us to go into the issue of the functions and resources of zilla parishads. It is only experience that could prove the merits of one type or another'.[1] It is therefore not possible to express any judgement on the finances of zilla parishad but it may be noted that from a very strictly financial point of view duplication of tiers of local bodies has not much to commend it and only renders the problem of matching local functions and local finances more difficult.

As an illustration of the resources and expenditure of the panchayati raj institutions we give in Table 13.4 the figures in respect of the state of Andhra Pradesh, which can be termed as a 'progressive' state in Panchayati-Raj matters. The table is self-explanatory.

[1]p. 33, Report of Study Team on Panchayati Raj Finances, 1963, Vol. I.

TABLE 13.4

Per Capita Income and Expenditure of Local Bodies in Andhra Pradesh (1962–63) (Rupees per capita)

A.Panchayats

Income		*Expenditure*	
House tax	0·34	Sanitation	0·16
Profession tax	0·03	Street lighting	0·10
Vehicle tax	0·03	Water supply	0·15
Tax on transfer of immovable property	0·37	Public utility works	0·47
Income from fees and fines	0·07	Miscellaneous expenditure	0·47
Others	0·73		
	1·57		1·35

B.Samithis

Normal Income	1·47	Education	0·61
Government grants	2·46	Medical and public health	0·14
		Roads	0·11
		Others	2·70
	3·93		3·56

C.Zilla Parishads

Government grants	3·09	Education	1·57
Others	1·57	Medical and public health	0·17
		Roads	0·62
		Others	2·02
	4·66		4·38

[Source: T. Rama Rao, 'Financing Panchayati Raj in Andhra Pradesh', *Asian Economic Review*, Feb., 1965, p. 217 and Report of Study Team on Panchayat Raj Finances, Vol. II.]

We may now briefly refer to some of the recommendations of the Study Team on Panchayati Raj Finances.

The Study Team found that frequent changes in the structure of functions and resources would prevent panchayati raj bodies from taking root. In other words some time should be given for the panchayati raj system to work before further charges or innovations are considered. The Study Team could muster little statistical data of a comprehensive type about the finances of the local bodies and has suggested the necessity for enquiries at the state level on a uniform basis by agreement among state governments, to be followed by an enquiry for the whole country instituted by the Government of India. More specifically, the Study Team considered the resources of each tier of Panchayati Raj institutions. In regard to panchayats, it recommended that house tax, profession tax and vehicle tax should be compulsory taxes in panchayats in all the states. Panchayats should also be given a share in land revenue as well as the surcharge on stamp duty and entertainment tax. For executing development projects, panchayats should have powers to levy special taxes on land revenue, house tax or on some other basis. All public lands, trees, tanks etc. should vest with the panchayats for exploitation. A matching grant should be given as an incentive to the levy and collection of panchayat taxes. A basic minimum maintenance grant of one Rupee per capita should also be given to every panchayat and this should be shared equally by the state and central governments.

As regards panchayat samithis, the Study Team recommended among others an annual per capita grant of Rupee One for each samithi to be earmarked for maintenance of staff on an agreed pattern. This grant should be shared equally by the state and central governments. It also suggested that an amount of Rs. 400 crores at Rs. 10 per capita of rural population should be allotted in the Fourth Plan for specified local developmental works to be given on a matching basis. The Study Team also stressed the need for avoidance of confusion and overlapping in taxation powers among various tiers for panchayati raj. Though it would not pronounce any definitive view on the role of zilla parishads, it considered that it was necessary to transfer some more elastic sources to zilla parishads in Maharashtra so that they could be an effective organ of local government at the district level. Another important recommendation of the Study Team was that each State should establish a Panchayati Raj Finance Corporation for the purpose of giving loans to panchayati raj bodies for various public utility undertakings.

The major difficulty of all the enquiries into local finances so far has been that there has been no attempt at a definite quantitative balance between the needs of the local bodies and their resources. The recommendations of the Study Team would have gained much more sense of urgency if they had been related to a comprehensive scheme for matching the functions and finances of local bodies. In the absence of such leading and constructive approach the recommendations have not received adequate recognition at the hands of the state governments.

The central problem of local finance is not so much the augmentation of revenue or the curtailment of expenditure as that of the purposeful matching of obligations and resources providing for the fulfilment of national priorities on the one hand and for a measure of local initiative on the other.[1] State governments, themselves in financial doldrums, are not likely to take any liberal view of the needs of local bodies nor hand over important sources of revenue to them. Our discussion can therefore centre on (1) ways to improve the existing devolutions to local bodies and (2) methods of augmentation of local incomes.

One of the difficulties which the local bodies have faced so far is not only that the grants given to them are meagre but that the provision for such grants was made in state budgets with reference to the financial position of the states from year to year and were not given in the form of a definite commitment for a specific period so that the local bodies could plan ahead. With the entrustment of plan schemes to local bodies they should have a firm idea of the resources they are likely to get for a plan period for the discharge of particular plan schemes. The result of an annual allocation was that the local bodies never looked farther than a year and the implementation of schemes was haphazard. Grants were also given by individual departments without any regard to the total position. If the states could complain of an unhealthy dualism in central assistance, the local bodies could very well complain of an indifferent pluralism in states' assistance!

Madras State took a step forward in this regard by framing what is known as the Panchayat Development Schematic Budget for the Third Plan period. What it did was to list in the form of a tabular statement, the resources of panchayat samithis[2] and panchayats, whether own resources, devolutions, grants or voluntary contributions from the people, for the entire plan period, separate columns being provided to show the resources of panchayats and samithis. Each panchayat samithi was asked to prepare a similar statement for its own area and detailed instructions were given regarding the manner of estimation of their resources, particularly grants. The result was that it was possible for the panchayat samithis to have an idea of the quantum of resources they would have for the plan

[1]See the author's 'Local Finance in Perspective' (Asia Publishing House, 1965).
[2]Designated as panchayat unions in Madras State.

o

period and they could therefore take advance and continuous action for the implementation of the schemes. They would also know the quantum of grants that would be available to them from year to year and the quantum of resources they might have to raise to utilize the grants. The framing of such a 'budget' likewise implied a willingness on the part of the state government to commit itself on the question of future quantum of grants to Panchayati Raj institutions but the cost of such commitment, if any, was far less than the advantage that the state government itself had a comprehensive picture of Panchayati Raj Finance which it never had before and knew where it stood and could reasonably be sure that the schemes it wanted to be implemented would be implemented smoothly. It would be true to say that the rather remarkable progress in the works programme in Madras (i.e. sinking drinking water wells, formation of roads and construction of school buildings) has been due to the willingness of the state government to commit itself fairly in advance to provide resources for this programme. The Study Team on Panchayati Raj Finances and the Study Group on Budgeting and Accounting Procedure in Panchayati Raj institutions[1] have commended the examination of this five year schematic budget for adoption by other states for this as well as other reasons.

Even the Panchayat Development Schematic Budget does not go far enough. It confines itself to the resources available but does not necessarily seem to match it to the responsibilities to be discharged. The proper course would be for governments to arrive at a comprehensive settlement of local finances once in five years, co-terminus with the Plan periods, with reference to certain desirable levels of expenditure. Unless such a settlement is made and the resources therefor committed, it may not be possible to expect the local bodies to function healthily and also carry out the various programmes of rural development that may be cast on them. In such a settlement grants will no doubt play an important part, because devolution of resources, particularly of land revenue, may have disequalizing effects. What is required of the states is not an abdication of their tax powers but a constructive and co-ordinated approach whereby the minimum necessary resources are given to local bodies in the best possible way which would ensure effective local government.

It should be admitted freely that there are no chances of local bodies discovering some spectacular new source of revenue. Nor will all state governments be keen on delegating their tax powers to the

[1]Government of India, 1963.

local bodies. In the circumstances, the search for increased local revenues has to be largely confined to the possibility of levy of surcharges on state taxes and to the question whether local bodies have exercised their existing tax powers well and whether they have collected whatever they levied. There is need for the gearing up of local bodies in the matter of their exercise of tax powers and collection of taxes. Experience shows that if the taxes could be related to local benefits there are greater chances of the local bodies willingly taxing the people.

This aspect becomes very relevant in the context of the proposal of state governments to abolish land revenue. The lines on which such abolition will proceed in various states are not quite clear. If land revenue has to be 'abolished' in some form or other, the proper course would be to make it a local tax. This will be worthwhile for a number of reasons: (1) The state governments can, if they like, reduce some of their grants to the local bodies so that the financial consequences of the abolition of land revenue are less severe to the states and the local bodies can be expected to levy their own rates for local improvements; (2) in any case a substantial portion of land revenue is even now given over to local bodies by devolution; (3) tying up local sacrifices and local benefits is the only way of broadening the local tax base; (4) object of taxing the rural sector, which is generally recognized to be undertaxed will be achieved.

Space would not permit us to discuss at length possible new sources of local revenue. It would however be necessary to mention that local bodies could augment their revenues by some minor local enterprises for which state governments can give loans. An institutionalization of this aspect through the establishment of a Panchayati Raj Finance Corporation in all the states was recommended by the Study Team on Panchayati Raj Finances. But it has not been implemented by any state so far, presumably because the Centre and the Reserve Bank are not coming forth for the contribution of their portion of the share capital.

A word is necessary about financial administration in local bodies. An ill-organized financial administration can be as much a handicap to local bodies as paucity of finances. The budgeting in local bodies requires far more care than has been exercised. Such budgets have also to be scrutinized carefully by higher authorities. There is need for greater care and personal responsibility on the part of executives in local bodies in matters relating to budgeting and accounting. A corps of well-trained accounting personnel is also necessary. The paucity of statistical data is another handicap which makes the states themselves rather unaware of what is really happening to local bodies. The statistical vacuum is there in spite of the fairly elaborate organization of Panchayati Raj institutions. This has to be remedied by well thought out action.

We may now refer to the finances of urban local bodies. The problem is of course the same, namely, the matching of resources and responsibilities. With growing industrialization as well as the popular awareness of the need for civic amenities, municipal functions have achieved a degree of importance which is not yet fully realized. The 'Report on the Augmentation of Financial Resources of Urban Local Bodies' (1965) by a Committee of Ministers constituted by the Central Council of Local Self-Government is the first full-blown report on municipal finances.[1]

This report has approached the problem of local finances more or less in the manner we have envisaged. It has taken pains to evolve certain standards of services for different levels of urban development. A table prepared by it, presented here as Table 13.5, shows clearly the resources gap if a balance between finances and functions at certain desirable levels of expenditure has to be reached. It will be readily seen from the table that the quantum of annual recurring requirements of the urban local bodies are of an order which the state governments can provide only with a heightened consciousness of the problems of urbanization.

Where the Committee has been less successful is on the methods of filling up the resources gap; nor does it seem to have made a clear distinction between current and capital needs. It suggested that the gap should be bridged (a) by better utilization of existing resources; (b) by assignment of a share or the entire proceeds of some state revenues; (c) by grants and (d) by new sources of revenue. The committee reiterated the recommendation of the Taxa-

[1]The Local Finance Enquiry Committee, 1951, and the Taxation Enquiry Commission 1953–54, also dealt with municipal finances but not with any particular result. Surprisingly enough, adequate statistical data even on municipal finances have been hard to come by. Mention should also be made of a preliminary survey of Municipal Finances in the Calcutta Metropolitan District by Abhijit Datta and David C. Romney (Asia Publishing House, 1963). As part of a series of studies of the Calcutta Metropolitan District it has special value; but it has not been able to go further than the collection and tabulation of the relevant financial statistics. A revealing insight into the costs of urbanisation can be had from an estimate of this study that the per capita state expenditure in West Bengal as a whole is one third of that of Calcutta itself and one-half of the per capita expenditure in the Calcutta Metropolitan District. As regards urbanisation in India in general, see 'India's Urban Future', ed. Roy Turner, (Oxford University Press, 1962).

TABLE 13.5

Gap between functions and finances of Urban local bodies

(Rs. Crores)

State (1)	Desired level of expenditure (2)	Present income (3)	Deficit (4)	(3) as percentage of (2) (5)
Andhra Pradesh	14·37	5·83	8·54	41
Assam	1·82	0·74	1·08	41
Bihar	8·21	3·43	4·78	42
Gujarat	14·74	10·70	4·04	73
Jammu and Kashmir	1·64	0·40	1·24	24
Kerala	4·64	1·45	3·19	31
Madhya Pradesh	10·87	5·40	5·47	50
Madras	18·58	11·85	6·73	64
Maharashtra	36·80	29·10	7·70	79
Mysore	13·53	6·99	6·54	52
Orissa	2·44	0·70	1·74	29
Punjab	9·95	6·94	3·01	70
Rajasthan	8·71	2·46	6·25	28
Uttar Pradesh	29·15	14·61	14·54	50
West Bengal	25·98	11·23	14·75	43
All India including Delhi	211·28	120·20	91·68	57

[Source: p. 37, Report on Augmentation of Financial Resources of Urban Local Bodies, 1965, Ministry of Health.]

tion Enquiry Commission for the reservation of certain taxes for exclusive utilization by or for local bodies. It also suggested that 25% of the proceeds of the motor vehicles tax should be earmarked for local bodies. These bodies should also have powers of levy of a surcharge or tax on the consumption of electricity. It also suggested that the urban local bodies should get a 'recurring annual per capita basic general purpose grant' at rates varying from 25 Paise in the case of Corporations to Rs. 1–50 in the case of small municipalities.

Considering the past attitude of state governments to local finances, the Committee could not probably have gone more specifically into methods of bridging the gap. It has undoubtedly blazed the trail for a purposive analysis of local finances by estimating the needs of local bodies quantitatively and by attempting to consider measures for bridging the gap. It has also classified urban local bodies into different categories and suggested differential treatment. The capital resources for urban local bodies, it has suggested, have to be obtained by them from an Urban Development Board to be established in each state which besides financing urban development would also enable the drawing up of master plans for various urban areas.

The Committee has however found the present position of municipal finances to be unsatisfactory. It found that at the existing revenues the urban local bodies cannot perform even their obligatory functions. It was itself unable to get requisite figures from all urban local bodies. It could get the data for only about 75% of the total number of urban bodies. This actually shows that at present there is no proper reporting system even for the urban local bodies. On examining the financial conditions of urban local bodies for the year 1960–61 the Committee came to the conclusion that whatever taxes local bodies are allowed today to levy are more or less the same which were exploited by them in Lord Ripon's time.[1] Further, in the absence of a clear cut demarcation between state and local sources of taxation there has been an encroachment by state governments in the legitimate domain of local taxes. Another disquieting feature of the present system of local finances, according to the committee, is that local bodies are increasingly deprived of their lucrative non-tax sources of revenue.

Table 13.6 taken from the Committee's report will show the per capita income and expenditure of urban local bodies in India during the year 1960–61.

Statewise figures would however show greater variations but it is not possible for us to analyze them in such detail here.

The length at which we have dealt with the problems of local finances in the context of the state finances has been due to the fact that state governments are willy nilly prone to neglect local finances and be content to take *ad hoc* and piece-meal decisions. When the state governments complain of the financial inequities of the central dispensation they are actually living in a glass house, because they are guilty to the same extent, if not more, when they deal with local bodies. No doubt the comparison will not be apt in every detail and local bodies have also to prove themselves as efficient administrative units. Nor should we ignore fact that the state finances themselves are in an unsatisfactory state when compared to their challenging responsibilities. At the same time our argument is not based on any doctrinaire views on local autonomy; nor can

[1]i.e. 1880's.

TABLE 13.6

Per Capita Analysis of Revenue and Expenditure of urban local bodies (1960–61) (Rupees)

State	Taxes	Ordinary grants	Total income	Total revenue expendi-ture	Public Health and conven-ience	Education	Public Works
Andhra Pradesh	7·53	2·30	12·15	10·26	4·99	2·22	1·08
Assam	4·70	3·39	9·94	5·74	3·47	0·17	0·37
Bihar	5·34	4·10	10·86	8·20	3·46	1·98	0·71
Gujarat	15·01	2·15	21·34	22·71	7·08	1·68	7·55
Jammu and Kashmir	6·02	0·03	7·24	7·97	3·34	0·05	1·79
Kerala	4·85	1·23	8·94	9·15	5·29	0·05	1·27
Madhya Pradesh	9·08	1·57	12·93	13·22	5·25	2·40	1·38
Madras	12·80	2·19	20·33	17·75	8·73	3·38	1·83
Maharashtra	20·74	1·84	27·21	24·70	11·89	3·40	1·26
Mysore	9·83	2·41	15·51	13·23	7·36	1·60	0·89
Orissa	3·04	3·05	7·25	8·35	3·78	0·94	1·87
Punjab	12·42	0·23	18·27	17·07	9·50	1·42	1·47
Rajasthan	4·70	1·00	7·60	6·20	3·50	0·20	0·30
Uttar Pradesh	7·28	2·55	13·49	12·03	4·83	1·93	1·06
West Bengal	9·86	2·29	15·12	14·39	6·34	0·82	0·84
All India including Delhi	10·61	2·32	17·23	16·20	7·47	2·19	1·62

[Source: pp. 379 and 383 of the Report on Augmentation of Financial Resources of Urban Local Bodies, 1966.]

it be taken as merely 'touching' the state governments for extra aid. Our call is only for consistent, rational and comprehensive thinking on state aid to local bodies because local autonomy on a starvation diet will be functionally self defeating.

In view of the importance of putting local functions and local finances in order, it is necessary that the union government should itself evince interest in this matter. By interest, we of course mean here, that it should part with a portion of resources so that they can be exclusively earmarked for the orderly functioning of local bodies based on a properly settled scheme of planned needs and resources.

CHAPTER XIV

CONCLUSION

In the foregoing chapters we have sought to spotlight the various aspects of state finances and have also indicated the spheres in which problems have arisen and solutions are necessary. In doing this we have refrained from an elaborate study of each of the aspects nor have we gone, in great detail, into the considerable statistical material we have presented in the tables. The reason is, our object has been to emphasize the perspectives and keep away from exercises which will confuse the scheme of the book. To this extent our study can be said to be exhaustive in the raising of issues but not exhaustive in treating them. Ours has been an endeavour which can be likened to a conducted tour of a great city like Rome in two days —an exercise which makes one familiar with all the landmarks and what is what but which is obviously inadequate for those keen to go into details. Our conducted tour would have served its purpose if the tourist resolves to come back to the subject with greater leisure and undiminished interest.

It is a pity that the various aspects of state finances except those relating to federal financial questions have not received adequate treatment. Problems of co-ordination among states as well as the necessity for inter-state research on taxation and other financial problems would underline the need for an inter-state research body on the subject. We have already suggested this in the previous chapters and we ought to reiterate the suggestion. The working of the Finance Commission once in five years no doubt ensures periodical interest in state finances but somehow, even in spite of the recommendations of the Finance Commissions themselves, there has been no adequate study of the financial problems of states. Many state governments do not even seem to have any cell for a study of their own problems. Decisions are taken in *ad hoc* manner and not necessarily based on a proper study of the facts relevant to the issues. Hence we should suggest that, apart from an inter-state body, each state government should establish in its Finance Department a study cell which can also do the useful job of collecting information in respect of other states and interpreting them.

CONCLUSION

It would have been noticed that we have not had a regular discussion on the comparative development levels of the states and as to how far central planning has resulted in an equal rate of growth in various states.[1] We have however referred to the problems of equalization in a number of chapters and have also produced some statistical material in this respect. In Chapter II we have seen in some detail the various indices of development of the states and in a later chapter we have touched on problems like equity of central aid. However, we have avoided a full dress debate on the question of the relative levels of development of various states. The reason for our not doing so is that such a discussion, unless done in great detail, is only likely to provide room for controversy. We have already mentioned that the problem of equal development of states cannot be approached in mechanistic terms. First of all, one has to select various indices of development for a comparison among states. In respect of some items of development one index may not be sufficient and one may have to take into account various factors. The historical administration of the state, its natural resources etc., also determine the level of development that it has the capacity to reach. The sense of priorities is also apt to vary from state to state. Some states, for example, may concentrate on education but some others may concentrate on health. Hence the problem of equalization and the role of central aid cannot be reduced to a simple question of quantifying development into various per capita expenditure levels, fixing some norms applicable to all states and working out the deficit for substandard states. While no doubt in practice some of the problems of equalization have to be tackled more or less in this fashion, the general question as such is not that simple. Even if such quantification is done and each state is given the money, its sense of priorities may be different; and if it is free to spend the money as it likes, each state would follow its own pattern. It is thus very difficult to arrange cardinally and even ordinally the various states in relation to their development. While it will no doubt be possible to pick out what are beyond doubt the better developed states and what are beyond doubt the less developed states, a general arrangement even in ordinal terms will be difficult. We must however point out that the question of equalization of development in the states has not been taken up in sufficient

[1]'The first decade of planning did not bring about any major reduction in the inter-state disparity in per capita income' according to a study undertaken by the National Council of Applied Economic Research, New Delhi. *The Mail*, 24th August, 1967.

detail. It has been announced that in the Fourth Plan central plan aid will be related to the population of the states. Population is no doubt a rough index of equity but it does not always work out correctly. Hence, only a comprehensive study, done without public debate, and with reference to various indices of development, can throw sufficient light on the problem.

Our study of state finances has revealed that they have now reached a stage where their condition should be termed as indicative of a fundamental disequilibrium, a situation which causes grave concern, if not alarm. We have tried to identify the reasons for this. We have found that the introduction of planning is the greatest single factor for the increase in expenditure of the states even though other factors on the non-development front have also been constantly pushing up expenditure. While individual states may have erred in their budgetary discipline from time to time it would be wrong to state that the present financial condition of the states is the result of states' extravagance. In recent years the cry of the 'extravagance' of the states has been raised from time to time. While such a criticism is no doubt desirable to keep erring states in check, the larger fact should not be forgotten that the main reason for the financial doldrums of many of the states has been the implementation of the plans.

This is not to say that plan schemes should not be implemented because they affect the finances of the states. That would be a counsel of despair. There is no doubt that most of the plan schemes have resulted in substantial benefits to the people in the fields of agriculture, education, health and other fields. A very large amount of capital expenditure has been incurred on the development of irrigation, electricity, road communications etc. These are substantial benefits.

But it is not as if no useful lessons can be drawn from the experience of the states in the past fifteen years. First of all, it has been seen that non-development expenditure has acquired a rate of growth of its own and the shift from non-development to development expenditure has been comparatively slow. But the causes for the growth of non-development expenditure do not always seem to be fully appreciated with the result that we hear more exhortations than concrete suggestions about cutting down non-development expenditure. One main reason for the growth of non-development expenditure is the mounting interest charges on the loans from central government. To this extent the capital account has necessarily its implications on the revenue account and this would call for a solution based on co-ordinated and thoughtful approach in planning and relating the revenue and capital accounts of the states.

If one tries to look into the future, the prospects are not bright and the current trends may continue unless steps are taken to halt them, if not reverse them. The shift from non-development to development expenditure will continue to be rather slow. At the same time the proportion of central aid in the finances of the states is likely to increase more and more. No doubt this proportion is kept in comparative check because of the increase in state revenues but even so the percentage of central aid will go on increasing.

In recent years many states have proved themselves to be capable of additional taxation. We have seen that there is considerable scope in taxation relating to sales tax and electricity duties. Even so a stage has now been reached when the implementation of new schemes has to be done not only by the increase of revenue but also by the rearrangement of expenditure by effecting savings in various respects. The problem of getting the best value for the money spent was never more urgent than it is today. There has to be a systematic, consistent and ruthless examination of all existing items of expenditure and such of those as are found to be unnecessary will have to give way to worthier schemes awaiting implementation. Doing this task is by no means easy and in doing it a certain boldness and a sense of proportion will be necessary. It should be possible to find out schemes both on the non-plan as well as the plan side which the state governments can do without. The Fourth Plan has done well by reckoning 'savings' (in administration) as an element of plan financing but it remains to be seen how far the anticipations are realized.

In analyzing the expenditure of the states it will also be necessary to give adequate emphasis to physical evaluation since this is an aspect which is too often neglected. This is best done at the time of scrutinizing budget estimates.

Our study has also shown that the borrowing patterns of the state governments would need change. We have suggested a composite solution to the debt problems of state governments. We have already indicated that the capital account of the state should not be quite out of proportion to the potentiality of the revenue account to service the debts. Every effort has to be made to make capital expenditure productive but it should not be forgotten that there are legitimate avenues of capital expenditure which are not productive at all. In this respect, the size and composition of the capital expenditure of the states has to be determined with much greater care. Our suggestion has been to make the state governments more conscious of the need to make their capital expenditure productive. One thing which we have touched in a previous chapter is the possibility of transferring the subject of electricity to the centre. This will take away a considerable portion of loan expenditure of state governments and thus reduce central control or at least the 'vertigo' of central assistance.[1] Equally importantly it will render possible the choice of schemes with reference to the greatest economic advantage whereas the present arrangement cannot redound to the choice of the most economic schemes of power generation. The states in a federation have to follow the policy of comparative advantage wherever it is possible. The difficulty is of course that no state would look at a proposal like this with any degree of warmth. The resistance of the states to such a proposal will have to be overcome sooner or later and it is time to begin to think in terms of a centralized system of power generation.

In the field of centre-state relations we have seen the phenomenal increase in central assistance. With such an increase in central assistance the basic federal financial equations of the distribution of revenues and through them the division of powers, become rather notional. We have seen how there is a controversial dualism in central assistance which it is high time to solve. We have also seen how though the quantum of plan aid has increased substantially, the centre gives it in a way which does not give satisfaction or elbow room to the states. The basic problem in planning is however not so much one of more grants-in-aid and less patterns. The problem is the real one of making the states active and enthusiastic partners in development. The present procedure of planning and plan financing do not make it easy for states to try out schemes on their own. Planning from below has become rather difficult.

Planning can be more effective if the states are encouraged to become capable of selecting really worthwhile schemes. It will also be necessary for this purpose to evolve a scheme whereby the experts of the centre who formulate various schemes are asked to come down to some state or other for a tenure of two or three years and implement the schemes they have formulated. This will render them more responsive to the difficulties of the field and the feelings

[1]The National Development Council decided in 1964 that some of the major multipurpose projects under the control of the states may be taken over by the centre so that the debts owned by the states may be transferred to itself. But this proposal has made no headway. 'Trends in States' Indebtedness', *Economic Times*, Bombay, 5th August, 1967.

of the states. Likewise people in the states who chaff at central policy formulation should be asked to go to the centre and try their hands in policy making. Such interflow between the centre and the states now exist only in respect of certain services but it should be made possible for technical services too, especially in fields like agriculture, community development, co-operation, animal husbandry, irrigation etc.

Our view of centre-state relations is therefore not merely one of solving a tricky problem in federal finance but one of securing that the states retain their freshness of thinking and their financial identity. The same holds good though in a somewhat restricted fashion in the sphere of state-local relations. We have said enough to point out the gross neglect of local finances. A complex of system of decentralization in the form of panchayati raj institutions has been put on the field but unless it is supported by adequate finances to make the decentralized setup work, whatever is spent is is not spent to the greatest advantage. A decentralized setup without the wherewithal to make it work will only create troublesome spots in the body politic without any commensurate return.

The results of the Fourth General Elections in 1967 have opened another chapter in the history of centre-state relations. The Congress Party has been returned to power in the centre but in as many as nine states non-Congress Governments have been formed. The future of centre-state relations in the context of different parties in power at the centre and in some of the states, has, therefore, become a live topic for debate. There is no doubt that the future pattern of centre-state relations will have substantial implications not only for state finances but for the economic development of the country. It is however neither possible nor necessary to make any immediate prognostications on the central-state relations of the future. We shall confine ourselves to making some general observations about the implications of the new situation for the subject of our study.

We may begin by noting the views of the new Chief Ministers. Almost all the Chief Ministers, whether belonging to the Congress Party or otherwise, have expressed their views. It has been common ground among them that (1) the states' autonomy should be preserved (or 'restored') and their operational freedom ensured; and (2) the governments of the states irrespective of their political complexion should co-operate with the central government in the interests of the country. The centre has likewise promised its cooperation with the state governments irrespective of their political complexion.[1]

When we are considering the implications of the new situation for the planning process of the country we should first understand how we consider the planning process to be simplified if the governments holding power in the states as well as the centre belong to the same party. Repeated demands for a larger flow of funds to the states can be taken as axiomatic on the part of any state government irrespective of its political complexion. The more specific areas of conflict or friction can be (1) on doctrinaire grounds, (2) because of the procedures and patterns of financial assistance, and (3) on the issue of the location of central projects. We may analyze briefly the possibilities in these areas of conflict.

It is not clear how far there will be centre-state differences on doctrinaire grounds in the immediate future. Except in Madras, the non-Congress governments have been coalition Governments which cannot ordinarily have or afford a keen doctrinaire urge. In any event it will take some time for them to forge an economic policy of their own. Hence, major issues like public sector and private sector, short term projects and long term projects, etc., may not immediately become live issues of debate demanding immediate solutions.

As regards the friction that can arise because of States' grievances about the procedures and patterns of financial assistance and planning and in the matter of the location of central projects, it is well to remember that conflicts on these points can arise and have arisen even when the same party was in power in both the centre and the states. There have been strong personalities as Chief Ministers in certain states who had the stature to have their views accepted by the centre without much ado. Besides, Chief Ministers have in recent years acquired an unparalleled position of power in the federal setup. From time to time, Chief Ministers belonging to the Congress Party have 'clashed' with the centre and its procedures and patterns of financial assistance. Hence, a measure of friction is part of the federal game and such an element will be there irrespective of the complexion of the political parties in power in the centre and the states. But there is an important difference, namely, that when the same party held power at both ends it was possible to settle some matters by a certain amount of outside

[1]For a summary of the views of the new Chief Ministers, Congress and non-Congress, on centre-state relations, see pp. 20–26 of *Link* (Bombay), 9th April, 1967.

negotiations and personal equations. Dead-locks and stalemates could be avoided by intimate and informal talks. But in the new setup such intimate and informal negotiations will not be possible in the same sense. To this extent the country is becoming truly federal and its capacity for working the federation successfully will be tested.

Since the Fourth Plan for states seems to be more or less finalized, matters of major import may not ordinarily be taken up *de novo* and to this extent one hopes there may be no occasion for a centre-state show-down. But the implications of the new situation for the central patterns of assistance are worth watching. Will the centre be able to simplify its patterns of assistance? Of course, under Article 282, the centre can refuse to assist a state if it declines to toe the line, but will the centre be in a position to do so? Will it be in the interest of the country and of the centre itself to reach a situation where the patterns become not the conduits of finance but bottlenecks? The patterns themselves are likely to be changed but it is not easy to say what exactly the changes would be. It is possible that the non-Congress Governments may, in the event of their not being successful in getting more unconditional grants, ask for a larger variety of patterns of assistance for schemes of their own. The dice seems to be equally loaded in favour of block grants on the one hand and more complicated patterns of assistance on the other.

It is also likely that the Finance Commission will become more important than what it is today. At least initially, non-Congress state Governments are likely to urge for more finances to be allocated to them through the Finance Commission. The Finance Commission will itself be in a position to take a bolder and freer view of its job in the context of the changed situation. Its role will become crucial and more difficult. It may have to face a number of ticklish issues. For example, if only some states scrap prohibition (four of them have done so after the elections), will the increased revenue be reckoned in full and financial assistance to those states reduced to that extent or will it be argued that since all states were not ready to scrap prohibition, the state which did scrap it as a measure of raising resources should not be penalized on that score? Similarly, if some states abolish land revenue and in different degrees, how should the loss in revenue be treated? Should it be compensated by extra financial assistance and if so to what extent?

In any event, the stage is set for more negotiations and consultations between the centre and the states and among the states themselves. The negotiations are not going to be easy. The states themselves may find it difficult to combine on some issues because their financial interests may differ. The need for institutional agencies for inter-governmental co-operation and consultation will become urgent and it is not unlikely that some formal or informal agency comes into being. Zonal Councils and other existing forums are bound to become more alive. But it is not institutionalization but the truly federal spirit that will be needed to settle centre-state issues which are relevant not only to state finances but to economic development as a whole. A spirit of give and take and a capacity for patience will be necessary on both sides. There is no need to be pessimistic about the future of centre-state relations or about states' finances in general. The country's capacity for working the federation will be put to test and there is no reason why it should not stand the test successfully.

The material necessary for a detailed analysis of the post-election budgets of state governments is not yet available. But the overall position is that except Orissa, Rajasthan, Uttar Pradesh and Gujarat all states have shown deficits of varying degrees.[1] The revenue deficit in quite a few states, particularly Andhra Pradesh, Bihar, Haryana and Madhya Pradesh is higher than what had been indicated in their interim budgets in March, 1967. In Kerala and Maharashtra, the revenue surpluses estimated in their earlier budgets have declined. The ways and means position of the state governments will therefore continue to be difficult. Some state governments have tried to reduce their expenditure by reducing the size of their annual plan but there have been other commitments which have made budget balancing a very difficult task indeed. Proposals for abolishing land revenue, the need to raise the Dearness Allowance of the employees to be in parity with those of the central government employees, expenditure on relief of drought and famine as well as steps taken to fulfil election pledges are all relevant factors in the situation.

Orissa was the only state which proposed additional taxation measures in the interim budget itself. Many other states have proposed additional taxation in their regular post-election budgets. The rates of sales tax on luxury goods and motor vehicles taxes have been increased. Rethinking on prohibition has also taken place and the actual withdrawal of prohibition in a few states will bring in a measure of increased revenue for them. Kerala has proposed to open up lotteries. It is to be hoped that in the next budget the state govern-

[1]This account is based on *Commerce*, 15th July 1967, p.109.

ments will be able to make further attempts to mobilize and conserve resources and prevent their financial position from worsening.

If one were to attempt to compress our arguments in this book into one single argument it would be that the central, state and local financial systems have to be integrated and treated in such a way that each can play the role expected of it. In this sense the problem of state finances is not one to be solved by sheer financial expertise or by accounting gymnastics or by discovering new sources of revenue. Ultimately the problem of state finances has to be treated alongside the problems of politics and public administration and with a general sense of proportion. It is such an outlook on the part of the politician and the administrator that can make it possible for them to come to grips with the problem which demands and will continue to demand attention in the national scene.

APPENDIX

Seventh Schedule of the Indian Constitution

(Article 246)

List I. Union List

1. Defence of India and every part thereof including preparation for defence and all such acts as may be conducive in times of war to its prosecution and after its termination to effective demobilization.
2. Naval, military and air forces; any other armed forces of the Union.
3. Delimitation of cantonment areas, local self-government in such areas, the constitution and powers within such areas of cantonment authorities and the regulation of house accommodation (including the control of rents) in such areas.
4. Naval, military and air force works.
5. Arms, firearms, ammunition and explosives.
6. Atomic energy and mineral resources necessary for its production.
7. Industries declared by Parliament by law to be necessary for the purpose of defence or for the prosecution of war.
8. Central Bureau of Intelligence and Investigation.
9. Preventive detention for reasons connected with Defence, Foreign Affairs, or the security of India; persons subjected to such detention.
10. Foreign Affairs; all matters which bring the Union into relation with any foreign country.
11. Diplomatic, consular and trade representation.
12. United Nations Organization.
13. Participation in international conferences, associations and other bodies and implementing of decisions made thereat.
14. Entering into treaties and agreements with foreign countries and implementing of treaties, agreements and conventions with foreign countries.
15. War and peace.
16. Foreign jurisdiction.

17. Citizenship, naturalization and aliens.
18. Extradition.
19. Admission into, and emigration and expulsion from India; passports and visas.
20. Pilgrimages to places outside India.
21. Piracies and crimes committed on the high seas or in the air; offences against the law of nations committed on land or the high seas or in the air.
22. Railways.
23. Highways declared by or under law made by Parliament to be national highways.
24. Shipping and navigation on inland waterways, declared by Parliament by law to be national waterways, as regards mechanically propelled vessels; the rule of the road on such waterways.
25. Maritime shipping and navigation, including shipping and navigation on tidal waters; provision of education and training for the mercantile marine and regulation of such education and training provided by States and other agencies.
26. Lighthouses, including lightships, beacons and other provision for the safety of shipping and aircraft.
27. Ports declared by or under law made by Parliament or existing law to be major ports, including their delimitation, and the constitution and powers of port authorities therein.
28. Port quarantine, including hospitals connected therewith; seamen's and marine hospitals.
29. Airways; aircraft and air navigation; provision of aerodromes; regulation and organization of air traffic and of aerodromes; provision for aeronautical education and training and regulation of such education and training provided by States and other agencies.
30. Carriage of passengers and goods by railway, sea or air, or by national waterways in mechanically propelled vessels.
31. Posts and telegraphs; telephones, wireless, broadcasting and other like forms of communication.
32. Property of the Union and the revenue therefrom, but as regards property situated in a State subject to legislation by the State, save in so far as Parliament by law otherwise provides.
33. Omitted.
34. Courts of wards for the estates of Rulers of Indian States.
35. Public debt of the Union.
36. Currency, coinage and legal tender; foreign exchange.
37. Foreign loans.
38. Reserve Bank of India.

39. Post Office Savings Bank.
40. Lotteries organized by the Government of India or the Government of a State.
41. Trade and commerce with foreign countries; import and export across customs frontiers; definition of customs frontiers.
42. Inter-State trade and commerce.
43. Incorporation, regulation and winding up of trading corporations, including banking, insurance and financial corporations but not including co-operative societies.
44. Incorporation, regulation and winding up of corporations, whether trading or not, with objects not confined to one State, but not including universities.
45. Banking.
46. Bills of exchange, cheques, promissory notes and other like instruments.
47. Insurance.
48. Stock exchanges and future markets.
49. Patents, inventions and designs; copyright; trademarks and merchandise marks.
50. Establishment of standards of quality for goods to be exported out of India or transported from one State to another.
52. Industries, the control of which by the Union is declared by Parliament by law to be expedient in the public interest.
53. Regulation and development of oilfields and mineral oil resources; petroleum and petroleum products; other liquids and substances declared by Parliament by law to be dangerously inflammable.
54. Regulation of mines and mineral development to the extent to which such regulation and development under the control of the Union is declared by Parliament by law to be expedient in the public interest.
55. Regulation of labour and safety in mines and oilfields.
56. Regulation and development of inter-State rivers and river valleys to the extent to which such regulation and development under the control of the Union is declared by Parliament by law to be expedient in the public interest.
57. Fishing and fisheries beyond territorial waters.
58. Manufacture, supply and distribution of salt by Union agencies; regulation and control of manufacture, supply and distribution of salt by other agencies.
59. Cultivation, manufacture, and sale for export, of opium.
60. Sanctioning of cinematograph films for exhibition.

61. Industrial disputes concerning Union employees.
62. The institution known at the commencement of this Constitution as the National Library, the Indian Museum, the Imperial War Museum, the Victoria Memorial and the Indian War Memorial, and any other like institution financed by the Government of India wholly or in part and declared by Parliament by law to be an institution of national importance.
63. The institutions known at the commencement of this Constitution as the Benares Hindu University, the Aligarh Muslim University and the Delhi University, and any other institution declared by Parliament by law to be an institution of national importance.
64. Institutions for scientific or technical education financed by the Government of India wholly or in part and declared by Parliament by law to be institutions of national importance.
65. Union agencies and institutions for—
   (a) professional, vocational or technical training, including the training of police officers; or
   (b) the promotion of special studies or research; or
   (c) scientific or technical assistance in the investigation or detection of crime.
66. Co-ordination and determination of standards in institutions for higher education or research and scientific and technical institutions.
67. Ancient and historical monuments and records, and archaeological sites and remains, declared by or under law made by Parliament to be of national importance.
68. The Survey of India, the Geological, Botanical, Zoological and Anthropological Surveys of India; Meteorological organizations.
69. Census.
70. Union public services; all-India services; Union Public Service Commission.
71. Union pensions, that is to say, pensions payable by the Government of India or out of the Consolidated Fund of India.
72. Elections to Parliament, to the Legislatures of States and to the offices of President and Vice-President; the Election Commission.
73. Salaries and allowances of members of Parliament, the Chairman and Deputy Chairman of the Council of States and the Speaker and Deputy Speaker of the House of the People.
74. Powers, privileges and immunities of each House of Parliament and of the members and the committees of each House; enforce-

ment of attendance of persons for giving evidence or producing documents before committees of Parliament or commissions appointed by Parliament.

75. Emoluments, allowances, privileges, and rights in respect of leave of absence, of the President and Governors; salaries and allowances, and rights in respect of leave of absence and other conditions of service of the Comptroller and Auditor-General.

76. Audit of the accounts of the Union and of the States.

77. Constitution, organization, jurisdiction and powers of the Supreme Court (including contempt of such Court), and the fees taken therein; persons entitled to practise before the Supreme Court.

78. Constitution and organization (including vacations) of the High Courts except provisions as to officers and servants of High Courts; persons entitled to practise before the High Courts.

79. Extension of the jurisdiction of a High Court and exclusion of the jurisdiction of any such High Court from, any Union territory.

80. Extension of the powers and jurisdiction of members of a police force belonging to any State to any area outside that State, but not so as to enable the police of one State to exercise powers and jurisdiction in any area outside that State without the consent of the Government of the State in which such area is situated; extension of the powers and jurisdiction of members of a police force belonging to any State to railway areas outside that State.

81. Inter-State migration; inter-State quarantine.

82. Taxes on income other than agricultural income.

83. Duties of customs including export duties.

84. Duties of excise on tobacco and other goods manufactured or produced in India except—

(a) alcoholic liquors for human consumption;

(b) opium, Indian hemp and other narcotic drugs and narcotics, but including medicinal and toilet preparations containing alcohol or any substance included in sub-paragraph (b) of this entry.

85. Corporation tax.

86. Taxes on the capital value of the assets, exclusive of agricultural land, of individuals and companies; taxes on the capital of companies.

87. Estate duty in respect of property other than agricultural land.

88. Duties in respect of succession to property other than agricultural land.

89. Terminal taxes on goods or passengers, carried by railway, sea or air; taxes on railway fares and freights.
90. Taxes other than stamp duties on transactions in stock exchanges and futures markets.
91. Rates of stamp duty in respect of bills of exchange, cheques, promissory notes, bills of lading, letters of credit, policies of insurance, transfer of shares, debentures, proxies and receipts.
92. Taxes on the sale or purchase of newspapers and on advertisements published therein.
92A. Taxes on the sale or purchase of goods other than newspapers where such sale or purchase takes place in the course of inter-State trade or commerce.
93. Offences against laws with respect to any of the matters in this List.
94. Inquiries, surveys and statistics for the purpose of any of the matters in this List.
95. Jurisdiction and powers of all courts, except the Supreme Court, with respect to any of the matters in this List; admiralty jurisdiction.
96. Fees in respect of any of the matters in this List, but not including fees taken in any court.
97. Any other matter not enumerated in List II or List III including any tax not mentioned in either of those Lists.

List II. State List

1. Public order (but not including the use of naval, military or air forces or any other armed forces of the Union in aid of the civil power).
2. Police, including railway and village police.
3. Administration of justice; constitution and organization of all courts, except the Supreme Court and the High Court; officers and servants of the High Court; procedure in rent and revenue courts; fees taken in all courts except the Supreme Court.
4. Prisons, reformatories, Borstal institutions and other institutions of a like nature, and persons detained therein; arrangements with other States for the use of prisons and other institutions.
5. Local government, that is to say, the constitution and powers of municipal corporations, improvement trusts, district boards, min-

ing settlement authorities and other local authorities for the purpose of local self-government or village administration.

6. Public health and sanitation; hospitals and dispensaries.

7. Pilgrimages, other than pilgrimages to places outside India.

8. Intoxicating liquors, that is to say, the production, manufacture, possession, transport, purchase and sale of intoxicating liquors.

9. Relief of the disabled and unemployable.

10. Burials and burial grounds; cremations and cremation grounds.

11. Education including universities, subject to the provisions of entries 63, 64, 65 and 66 of List I and entry 25 of List III.

12. Libraries, museums and other similar institutions controlled or financed by the State; ancient and historical monuments and records other than those declared by or under law made by Parliament to be of national importance.

13. Communications, that is to say, roads, bridges, ferries, and other means of communication not specified in List I; municipal tramways; ropeways; inland waterways and traffic thereon subject to the provisions of List I and List III with regard to such waterways; vehicles other than mechanically propelled vehicles.

14. Agriculture, including agricultural education and research protection against pests and prevention of plant diseases.

15. Preservation, protection and improvement of stock and prevention of animal diseases; veterinary training and practice.

16. Pounds and the prevention of cattle trespass.

17. Water, that is to say, water supplies, irrigation and canals, drainage and embankments, water storage and water power subject to the provisions of entry 56 of List I.

18. Land, that is to say, rights in or over land, land tenures including the relation of landlord and tenant, and the collection of rents; and agricultural loans; colonization.

19. Forests.

20. Protection of wild animals and birds.

21. Fisheries.

22. Courts of wards subject to the provisions of entry 34 of List I; encumbered and attached estates.

23. Regulations of mines and mineral development subject to the provisions of List I with respect to regulation and development under the control of the Union.

24. Industries subject to the provisions of entries 7 and 52 of List I.

25. Gas and gas-works.
26. Trade and commerce within the State subject to the provisions of entry 33 of List III.
27. Production, supply and distribution of goods subject to the provisions of entry 33 of List III.
28. Markets and fairs.
29. Weights and measures except establishment of standards.
30. Money-lending and money-lenders; relief of agricultural indebtedness.
31. Inns and inn-keepers.
32. Incorporation, regulation and winding up of corporations, other than those specified in List I, and universities; unincorporated trading, literary, scientific, religious and other societies and associations; co-operative societies.
33. Theatres and dramatic performances; cinemas subject to the provisions of entry 60 of List I; sports, entertainments and amusements.
34. Betting and gambling.
35. Works, lands and buildings vested in or in the possession of the State.
36. Omitted.
37. Elections to the Legislature of the State subject to the provisions of any law made by Parliament.
38. Salaries and allowances of members of the Legislature of the State, of the Speaker and Deputy Speaker of the Legislative Assembly and, if there is a Legislative Council, of the Chairman and Deputy Chairman thereof.
39. Powers, privileges and immunities of the Legislative Assembly and of the members and the committees thereof, and, if there is a Legislative Council, of that Council and of the members and the commitees thereof; enforcement of attendance of persons for giving evidence or producing documents before committees of the Legislature of the State.
40. Salaries and allowances of Ministers for the State.
41. State public services; State Public Service Commission.
42. State pensions, that is to say, pensions payable by the State or out of the Consolidated Fund of the State.
43. Public Debt of the State.
44. Treasure trove.
45. Land revenue, including the assessment and collection of revenue, the maintenance of land records, survey for revenue purposes and records of rights, and alienation of revenues.

46. Taxes on agricultural income.
47. Duties in respect of succession to agricultural land.
48. Estate duty in respect of agricultural land.
49. Taxes on lands and buildings.
50. Taxes on mineral rights subject to any limitations imposed by Parliament by law relating to mineral development.
51. Duties of excise on the following goods manufactured or produced in the State and countervailing duties at the same or lower rates on similar goods manufactured or produced elsewhere in India—
   (a) alcoholic liquors for human consumption;
   (b) opium, Indian hemp and other narcotic drugs and narcotics; but not including medicinal and toilet preparations containing alcohol or any substance included in sub-paragraph (b) of this entry.
52. Taxes on the entry of goods into a local area for consumption, use or sale therein.
53. Taxes on the consumption or sale of electricity.
54. Taxes on the sale or purchase of goods other than newspapers, subject to the provisions of entry 92A of List I.
55. Taxes on advertisements other than advertisements published in the newspapers.
56. Taxes on goods and passengers carried by road or on inland waterways.
57. Taxes on vehicles, whether mechanically propelled or not, suitable for use on roads, including tramcars subject to the provision of entry 35 of List III.
58. Taxes on animals and boats.
59. Tolls.
60. Taxes on professions, trades, callings and employments.
61. Capitation taxes.
62. Taxes on luxuries, including taxes on entertainments, amusements, betting and gambling.
63. Rates of stamp duty in respect of documents other than those specified in the provision of List I with regard to rates of stamp duty.
64. Offences against laws with respect to any of the matters in this List.
65. Jurisdiction and powers of all courts, except the Supreme Court, with respect to any of the matters in this List.
66. Fees in respect of any of the matters in this List, but not including fees taken in any court.

LIST III. CONCURRENT LIST

1. Criminal law, including all matters included in the Indian Penal Code at the commencement of this Constitution but excluding offences against laws with respect to any of the matters specified in List I or List II and excluding the use of naval, military or air forces or any other armed forces of the Union in aid of the civil power.
2. Criminal procedure, including all matters included in the Code of Criminal Procedure at the commencement of this Constitution.
3. Preventive detention for reasons connected with the security of a State, the maintenance of public order, or the maintenance of supplies and services essential to the community; persons subjected to such detention.
4. Removal from one State to another State of prisoners, accused persons and persons subjected to preventive detention for reasons specified in entry 3 of this List.
5. Marriage and divorce; infants and minors; adoption; wills, intestacy and succession; joint family and partition; all matters in respect of which parties in judicial proceedings were immediately before the commencement of this Constitution subject to their personal law.
6. Transfer of property other than agricultural land; registration of deeds and documents.
7. Contracts, including partnership, agency, contracts of carriage, and other special forms of contracts, but not including contracts relating to agricultural land.
8. Actionable wrongs.
9. Bankruptcy and insolvency.
10. Trust and Trustees.
11. Administrators-General and official trustees.
12. Evidence and oaths; recognition of laws, public acts and records, and judicial proceedings.
13. Civil procedure, including all matters included in the Code of Civil Procedure at the commencement of this Constitution, limitation and arbitration.
14. Contempt of court, but not including contempt of the Supreme Court.
15. Vagrancy; nomadic and migratory tribes.
16. Lunacy and mental deficiency, including places for the reception or treatment of lunatics and mental deficients.

17. Prevention of cruelty to animals.
18. Adulteration of foodstuffs and other goods.
19. Drugs and poisons, subject to the provisions of entry 59 of List I with respect to opium.
20. Economic and social planning.
21. Commercial and industrial monopolies, combines and trusts.
22. Trade Unions; industrial and labour disputes.
23. Social security and social insurance; employment and unemployment.
24. Welfare of labour including conditions of work, provident funds, employers' liability, workmen's compensation, invalidity and old age pensions and maternity benefits.
25. Vocational and technical training of labour.
26. Legal, medical and other professions.
27. Relief and rehabilitation of persons displaced from their original place of residence by reason of the setting up of the Dominions of India and Pakistan.
28. Charities and charitable institutions, charitable and religious endowments and religious institutions.
29. Prevention of the extension from one State to another of infectious or contagious diseases or pests affecting men, animals or plants.
30. Vital statistics including registration of births and deaths.
31. Ports other than those declared by or under law made by Parliament or existing law to be major ports.
32. Shipping and navigation on inland waterways as regards mechanically propelled vessels, and the rule of the road on such waterways, and the carriage of passengers and goods on inland waterways subject to the provision of List I with respect to national waterways.
33. Trade and Commerce in, and the production, supply and distribution of—
   (a) the products of any industry where the control of such industry by the Union is declared by Parliament by law to be expedient in the public interest, and imported goods of the same kind as such products;
   (b) foodstuffs, including edible oil-seeds and oils;
   (c) cattle fodder, including oil-cakes and other concentrates;
   (d) raw cotton, whether ginned or un-ginned, and cotton seed; and
   (e) raw jute.
34. Price control.

35. Mechanically propelled vehicles including the principles on which taxes on such vehicles are to be levied.
36. Factories.
37. Boilers.
38. Electricity.
39. Newspapers, books and printing presses.
40. Archaeological sites and remains other than those declared by or under law made by Parliament to be of national importance.
41. Custody, management and disposal of property (including agricultural land) declared by law to be evacuee property.
42. Acquisition and requisitioning of property.
43. Recovery in a State of claims in respect of taxes and other public demands, including arrears of land-revenue and sums recoverable as such arrears, arising outside that State.
44. Stamp duties other than duties or fees collected by means of judicial stamps, but not including rates of stamp duty.
45. Inquiries and statistics for the purposes of any of the matters specified in List II or List III.
46. Jurisdiction and powers of all courts, except in Supreme Court, with respect to any of the matters in this List.
47. Fees in respect of any of the matters in this List, but not including fees taken in any court.

BIBLIOGRAPHY

P. P. Agarwal: *System of Grants-in-aid in India.* (Asia)

S. P. Aiyar and Usha Metha, ed.: *Essays in Indian Federalism.* (Asia)

Granville Austin: The *Indian Constitution, Cornerstone of a Nation.* (Clarendon)

R. N. Bhargava: *Theory and Working of Union Finance in India.* (Allen and Unwin)

R. N. Bhargava: *Indian Public Finances.* (Allen and Unwin)

B. R. Braibanti and J. J. Spengler: *Administration and Economic Development in India.* (Cambridge University Press)

A. H. Birch: *Federalism, Finance and Social Legislation.* (Oxford)

Asok Chanda: *Federalism in India.* (Allen and Unwin)

R. J. Chelliah: *Basis of Taxation in the Context of the Developing Indian Economy.* (Popular Prakashan, Bombay)

D. R. Gadgil: *Planning and Economic Policy in India.* (Asia)

Goswami: *Economic Development of Assam.* (Asia)

A. H. Hansen and H. S. Perloff: *State and Local Finance in the National Economy.* (W. W. Norton & Co.)

A. H. Hanson: *The Process of Planning.* (Oxford University Press)

U. K. Hicks et al.: *Federalism and Economic Growth.* (Allen and Unwin)

Indian Merchants' Chamber: *Union Budgets, a factual study of the finances of the Government of India, 1950–51 to 1964–65.*

Dibakar Jha: *Bihar Finances, 1912–13—1960–61.* (Granthmala Karyalaya, Patna)

A. Krishnaswamy: *The Indian Union and the States—A study in autonomy and integration.* (Pergamon Press)

J. A. Maxwell: *Federal Subsidies to the Provincial Government in Canada.*

B. R. Misra: *Economic Aspects of the Indian Constitution.* (Orient Longmans)

B. R. Misra: *Indian Federal Finance.* (Orient Longmans)

National Council of Applied Economic Research: *Economic Classification of Central and State Government Budgets.*

National Council of Applied Economic Research: *Distribution of National Income by States.*

A. Premchand: *Control of Public Expenditure in India.* (Allied Publishers)

Punjab University: *Central and State Government Budgets in India, an economic classification.*

V. V. Ramanadham: *The Economy of Andhra Pradesh.* (Asia)

BIBLIOGRAPHY

B. U. Ratchford: *Public Expenditure in Australia.* (Duke University Press)

Report of the National Bureau of Economic Research: *Public Finances, Needs, Sources and Utilisation.* (Princeton)

K. Santhanam: *Union-State Relations in India.* (Asia)

K. Santhanam: *Transition in India.* (Asia)

K. V. S. Sastri: *Federal-State Fiscal Relations in India.* (Oxford University Press)

N. V. Sovani and V. M. Dandekar: *Changing India.* (Asia)

R. N. Tripathy: *Federal Finance in a Developing Economy.* (World Press, Calcutta)

K. Venkataraman: *Local Finance in Perspective.* (Asia)

The Five Year Plans of India.

Reports of the Finance Commissions.

Reserve Bank of India Bulletins.

Combined Finance and Revenue Accounts of the Central and State Governments, Comptroller and Auditor-General of India.

Report of the Taxation Enquiry Commission, 1953–54.

Budgets of State Governments.

Memorandum for the Fourth Finance Commission, Government of Madras.

Report on the Augmentation of financial resources of Urban Local Bodies.

Report of the Bombay Sales Tax Enquiry Committee, 1957–58.

Report of the Committee on Transport Policy and Coordination, 1966.

Report of the Committee on the Working of State Electricity Boards, 1965.

Report of the Local Finance Enquiry Committee.

Report of the Rajasthan Finance Enquiry Committee.

Report of the Study Team on Panchayati Raj Finances.

Note: This bibliography does not claim to be comprehensive. Omission to include any work reflects no judgment on it! The sole idea is to furnish a list of works which, according to the author, have relevance to the broad framework in which states' finances have to be studied. Papers in periodicals have not been listed.

INDEX

Note: Statistical material has not been indexed. See List of Tables.

For Product Safety Concerns and Information please contact our EU representative GPSR@taylorandfrancis.com Taylor & Francis Verlag GmbH, Kaufingerstraße 24, 80331 München, Germany